12th International Conference on Environment Pollution and Prevention (ICEPP 2024)

E3S Web of Conferences Volume 620 (2025)

Brisbane, Australia
8-10 November 2024

Editor:

R. J. Haynes

ISBN: 979-8-3313-1552-8

Printed from e-media with permission by:

Curran Associates, Inc.
57 Morehouse Lane
Red Hook, NY 12571

Some format issues inherent in the e-media version may also appear in this print version.

For additional information, please contact EDP Sciences – Web of Conferences at the address below.

EDP Sciences – Web of Conferences
17, Avenue du Hoggar
Parc d'Activité de Courtabœuf
BP 112
F-91944 Les Ulis Cedex A
France

Phone: +33 (0) 1 69 18 75 75
Fax: +33 (0) 1 69 28 84 91

contact-edps@webofconferences.org

Additional copies of this publication are available from:

Curran Associates, Inc.
57 Morehouse Lane
Red Hook, NY 12571 USA
Phone: 845-758-0400
Fax: 845-758-2633
Email: curran@proceedings.com
Web: www.proceedings.com

TABLE OF CONTENTS

Committees

Conference Chairs

Sunil Herat, Griffith University, Australia
R. J. (Dick) Haynes, The University of Queensland, Australia

Conference Co-Chairs

Xingbin Sun, Northeast Forestry University, China
Dennis Y.C. Leung, University of Hong Kong, Hong Kong, China

Program Chairs

Rouzbeh Abbassi, Macquarie University, Sydney, Australia
Sue Charlesworth, Coventry University, United Kingdom
Roland Kallenborn, Norwegian University of Life Sciences, Norway

Technical Committee

Chun-Hung(Jeb) Lee, Dean, Center for Sustainable Development, National DongHwa University, Taiwan
Ghani Albaali, Princess Sumaya University for Technology PSUT, Jordan
Yubao Wang, Xinjiang University;Xi'an Jiaotong University, China
Gargh Shankar, Environmental Disasters Research Institute, India
Alaa El-Din Hamid Sayed, Assiut University, Egypt
Bangalore Suryanarayanaiah Rajanikanth, Indian Institute of Science, India
Seongjin Hong, Chungnam National University, Korea
Neeru Bansal, CEPT University, India
Rajeev Kumar Mishra, Delhi Technological University,India
Angela T Ragusa, Charles Sturt University, NSW Australia
Md Kamrul Hasan, Zhejiang University, China
Saikat Sinha Ray, Ulsan National Institute of Science and Technology, South Korea
Fu-Ming Chang, Da-Yeh University, Taiwan, China
Dinesh K. Yadav, ICAR-National Academy of Agricultural Research Management, India
Meisam Ahmadi Ghadikolaei, The Hong Kong University of Science and Technology, Hong Kong, China
Md. Mizanur Rahman, University Technology Malaysia, Malaysia
Mylene G. Dinglasan, Philippine Christian University, Philippine
Leena Ajit Kaushal, Management Development Institute Gurgaon, India
Linhui Zeng, Hangzhou XiHu District City Management and Security Service Center,

Statement of Peer review

In submitting conference proceedings to E3S Web of Conferences, I certify to the Publisher that I adhere to the **Policy on Publishing Integrity** of the journal in order to safeguard good scientific practice in publishing.

1. All articles have been subjected to peer review administered by the proceedings editors.
2. Reviews have been conducted by expert referees, who have been requested to provide unbiased and constructive comments aimed, whenever possible, at improving the work.
3. Proceedings editors have taken all reasonable steps to ensure the quality of the materials they publish and their decision to accept or reject a paper for publication has been based only on the merits of the work and the relevance to the journal.

Title, date and place of the conference

2024 12th International Conference on Environment Pollution and Prevention (ICEPP 2024)
Brisbane, Australia, 8-10 November, 2024

Proceedings editor(s): R. J. (Dick) Haynes

Date and editor's signature 2025.3.5

Preface

2024 12th International Conference on Environment Pollution and Prevention (ICEPP 2024) was successfully held during 8-10 November, 2024 in Brisbane, Australia. ICEPP conference is an annual event which has been conducted since 2013 and will be held every year. The aim of the conference is to bring together innovative academics and industrial experts in the field of Environment Pollution and Prevention to a common forum. It is a great program at this conference that allowed presenters to reflect past accomplishments, renew friendships and extend our networks, and jointly explore current and future research directions. We believe that participants had a productive and fun-filled time at this very special conference.

ICEPP2024 features four distinguished keynote speakers, 2 onsite sessions one poster session and one online session with different topics: Wastewater Treatment and Air Pollution Control; Waste Management and Environmental Sustainability; Environmental Pollution Control and Sustainable Development Strategies. Participants came from various countries including: Australia, Hong Kong, Jordan, China, Thailand, Taiwan, Vietnam, Korea, India, Indonesia, Canada, Spain etc. It serves as an excellent opportunity to discuss the latest innovations, trends, concerns, practical challenges, and solutions in the related fields.

The papers included in the proceedings are a compilation of the accepted papers and represent the main outcome of the conference. Each submission underwent rigorous peer review to ensure that only the most outstanding and innovative research results are included in this proceeding

Thank you to all participants, sponsors, and supporters whose enthusiasm and commitment have made ICEPP 2024 a resounding success. Hopefully, all participants and other interested readers will benefit scientifically from the proceedings

Conference Chair
R. J. (Dick) Haynes,
The University of Queensland, Australia

Exposing privacy risks in indoor air pollution monitoring systems

Krishna Singh*, *Shreyash* Gujar, *Sachin* Chaudhari, and *Ponnurangam* Kumaraguru

International Institute of Information Technology, Hyderabad, India

Abstract. Indoor air pollution monitoring has been a region of interest in recent times. Multiple Internet of Things (IoT) enabled devices are available for this purpose. With the growing number of sensors in our daily environment, huge amounts of data are being collected and pushed to the servers through the Internet. This study aims to reveal that seemingly trivial indoor air pollution data containing particulate matter, carbon dioxide, and temperature can reveal complex insights about an individual's lifestyle. Data was collected over a period of four months in a real-world environment. The study demonstrates the inference of cooking activities by using machine learning and deep learning techniques. The study further demonstrates that different food items and culinary practices have different air pollution signatures, which can be identified and distinguished with great accuracy (>90%). In the practice of inferential analysis, it is not necessary to rely on data characterised by high frequency or granularity. Less detailed data like hourly averages, can be used to make meaningful conclusions that might intrude on an individual's privacy. With the rapid advancement in machine learning and deep learning, a proactive approach to privacy is needed to ensure that the collected data and its usage do not intentionally or unintentionally breach individual privacy.

Keywords: Air pollution, Cooking activities, Dataset, Deep learning, IoT, Machine learning, Privacy

1 Introduction

Indoor air conditions significantly impact an individual's health, leading to an estimated 3.2 million deaths worldwide annually in 2020 [1]. Monitoring indoor air pollution has become a prominent focus for urban populations. As industrialisation and urbanisation continue to escalate, indoor air quality has garnered public concern and holds excellent importance [2]. Various IoT devices, including portable sensors and comprehensive smart home systems, are available in the market, enabling tracking of multiple pollutants within indoor environments.

The IoT has unleashed significant potential in the computing domain, with an estimated growth rate of 10.77% per annum until 2032 [3]. The rapid increase in consumer IoT devices, which incorporate a growing number of sensors into our everyday surroundings, is a crucial driver of this growth. These IoT devices have become essential in people's daily lives, from smart home monitoring and automation to smart wearables. Any physical object can be transformed into a smart device, enabling large-scale data collection and communication over

* Corresponding author: krishna.singh@research.iiit.ac.in

the Internet. However, we argue that the data collected by these devices can pose a risk to individuals' privacy by considering the use case of indoor air pollution monitoring.

Several studies have previously highlighted the privacy implications of IoT. According to [4], Multi-variate sensor nodes that measure temperature, illumination, sound, CO2, and passive infrared (PIR) can be used to infer the number of occupants. Moreover, [5] demonstrates that smart electric meters, which record and transmit energy consumption data at regular intervals, can reveal insights into household activities and appliance usage patterns. In addition, the work presented in [6] illustrates how data from IoT sensors can be leveraged for localisation, activity recognition, and user input inference. [7] uncovers the privacy implications of Smart TVs, suggesting that they function as remote sensors for a cloud-based data center. [8] illustrates the use of commercial sensor in homes to infer cooking activity timings to support food journalism.

Through this work, we aim to demonstrate how indoor air pollution monitoring in homes can reveal complex insights about an individual's living patterns and pose significant risks to privacy by monitoring basic air quality parameters: PM2.5, PM10 (Particulate Matter), CO_2 (Carbon Dioxide) and Temperature. First, we use machine learning and deep learning algorithms to demonstrate the inference of cooking activity timings from the data. The inferences were improved by monitoring the outside air pollution data alongside.

We then demonstrate that different food items and preparation methods have different air pollution data signatures, which can be inferred from the data using machine learning algorithms, causing a significant breach of an individual's privacy. For this, we chose the controlled environment of Kadamb Mess cafeteria kitchen at the International Institute of Information Technology, Hyderabad (IIIT-H). The menu for each day of the week is fixed for four meals: Breakfast, Lunch, Evening snacks and Dinner. This controlled setup allows us to collect and analyse air pollution data corresponding to specific food items and cooking methods associated with each day. By successfully identifying the day of the week based solely on the pollution data, we effectively demonstrate the ability to distinguish between the unique air pollution signatures of various culinary preparations, thereby highlighting significant privacy concerns.

2 Measurement setup

2.1 Sensor Node

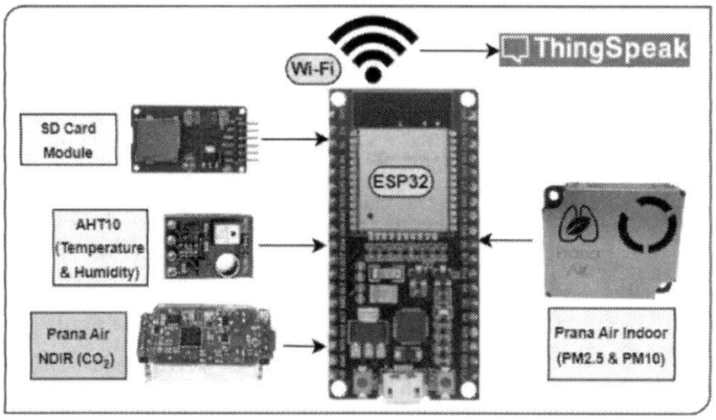

Fig. 1. Hardware interfacing diagram showing the data transfer directions with arrows.

The hardware setup comprises three sensors interfaced with a Wi-Fi-enabled ESP32 microcontroller. The setup as depicted in Fig. 1 is placed in an IP65 box with air intakes to facilitate the proper flow of air. AHT10 High Precision Digital sensor is used to measure temperature and humidity, Prana Air Indoor PM sensor [9] is utilised for PM2.5 and PM10 measurement, and Prana Air CO2 NDIR [10] sensor is used for CO_2 measurement. The choice of the sensors was based on their cost-effectiveness. The specifications of each sensor are mentioned in Table 1. The ESP32 module connects to a local Wi-Fi network, transmitting the sensor data to a ThingSpeak channel. ThingSpeak is an open-source IoT platform by MathWorks facilitating real-time monitoring and storage of data from sensors for further analysis. An SD Card Module is interfaced to prevent data loss due to network connectivity issues, which stores data in a CSV file.

Table 1. Performance metrics of sensors validate their choice due to good range and accuracy.

	Parameter Sensed	Range	Accuracy	Resolution
AHT10	Temperature	-40°C to 85°C	±0.3°C	0.01°C
	Humidity	0% to 100%	±2%	0.024%
Prana Air CO2	CO_2	1-10000 ppm	±2%	1 ppm
Prana Air Indoor PM	PM2.5	0-1500 µg/m³	±10% or ±10 µg	1 µg/m³
	PM10	0-2000 µg/m³	±15% or ±15 µg	1 µg/m³

2.2 Calibration

Sensors used for data collection are cost-effective, unlike the commercial sensor node used in [8]. These sensors are not very accurate when used as is but work well when calibrated with an accurate reference sensor as shown in [11] for the CO_2 sensor. Thus, we perform calibration with the help of a reference sensor, Aeroqual S500 [12]. The methodology followed is the same as described in [13]. The sensor nodes and reference sensor were kept in a controlled environment for three days. Simple linear regression was used to adjust for the differences in measured values of PM2.5, PM10, and CO2 between the sensor nodes and the reference sensor for each sensor node individually. Fig. 2 shows the sensor node and the Aeroqual S500 reference sensor.

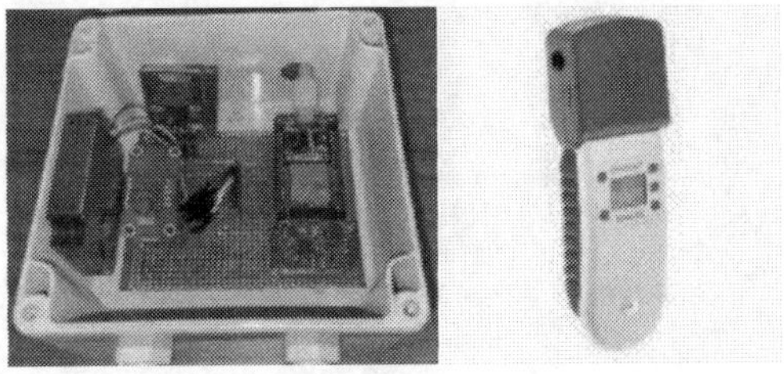

Fig. 2. Sensor node (left) and reference node (right).

2.3 Data collection

Four nodes were deployed in the cooking area of the cafeteria of the IIIT-H. A detailed map of the cafeteria kitchen and node locations is shown in Fig. 3. Red-colored stations correspond to stove cooking, blue-colored stations correspond to steam cooking, and green-colored stations denote the areas where preparations for cooking take place. The nodes were placed at sufficient distances from the stoves to ensure the operating temperatures did not affect the sensor values.

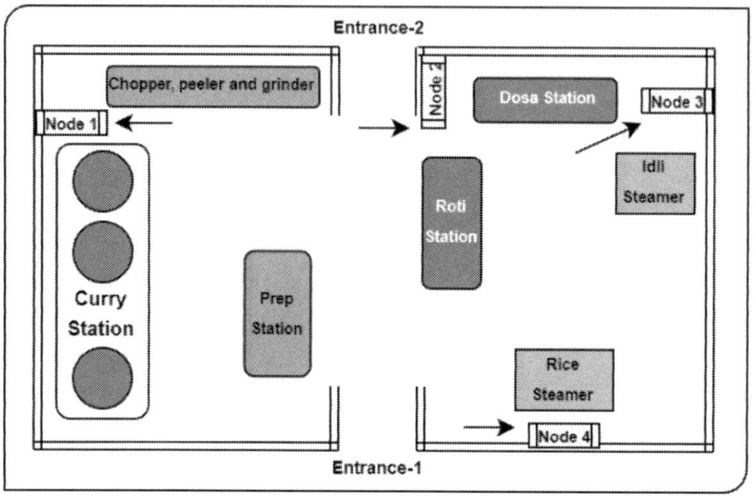

Fig. 3. Node deployment map with placement of nodes and cooking stations for various food items.

Data was collected over a period of four months, from January 2024 to April 2024. The sensing frequency was 30 seconds. Node-1 was placed near the Curry Station to capture the impact of slow and prolonged cooking. Node-2 was placed near Roti Station and Dosa (Indian thin pancake) Station, capturing the effect of rapid and high-flame cooking. Node-3 was near the Dosa Station and Idli (dunken rice cake) Steamer to capture the impact of both. Finally, Node-4 was placed near the Rice Steamer to capture the effect of steam cooking. Fig. 4 shows a sample time series plot of the collected data and binary labels 0 and 1, denoting "Not-Cooking" and "Cooking" activities, respectively. Variations in different parameters can be clearly observed during cooking activities.

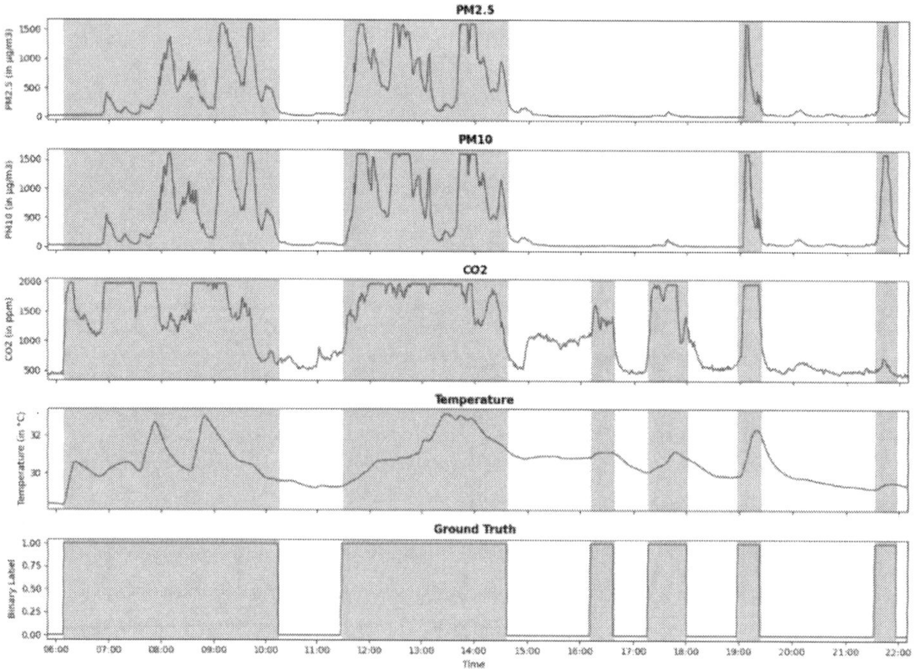

Fig. 4. Sample time series plot for a day from 6 am to 10 pm.

3 Methodology

Information gathered from the nodes was used to create datasets for classification purposes. Fig. 5 provides an overview of the dataset creation process. The data underwent preprocessing to ensure accuracy. Relevant features related to the specific task were identified and extracted from the data. These features were normalised, and appropriate class labels were assigned to each feature vector to enable future classification tasks. Various machine learning and deep learning models were then utilised to determine the most suitable model for the task.

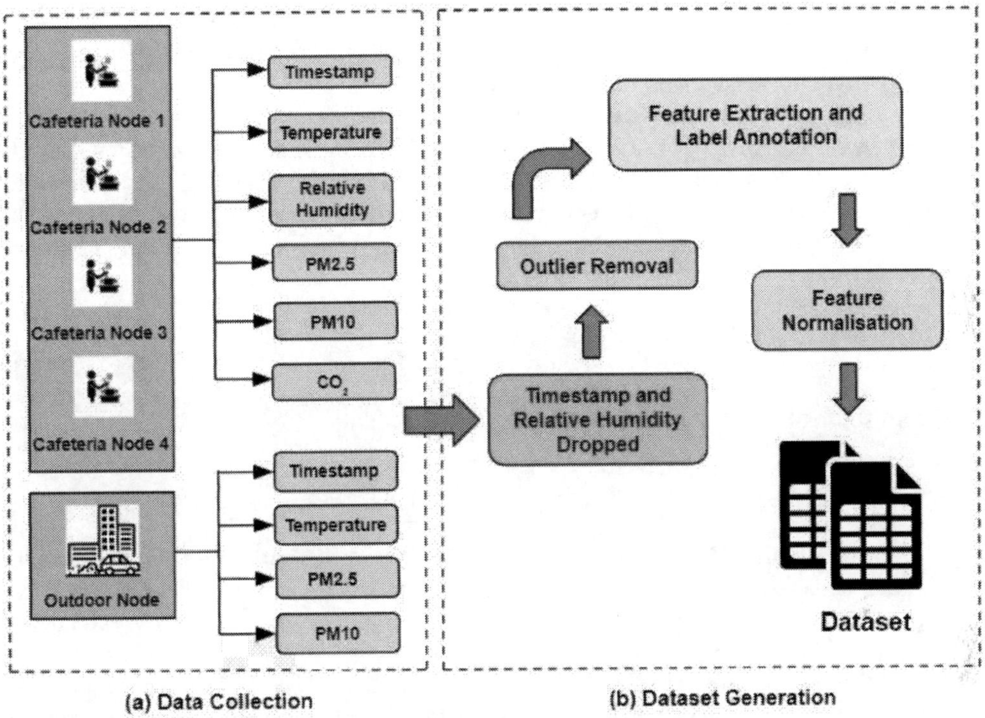

Fig. 5. Data flow diagram showing the outline of data processing from nodes to dataset.

3.1 Pre-Processing

During the data analysis, it was identified that humidity had a minimal correlation and was subsequently excluded. The timestamp was replaced by a unique "Entry ID" to mitigate temporal influence on predictions. Outlier values were eliminated using thresholding, and the features were normalised using standard normalisation.

3.2 Binary classification of cooking activity

3.2.1 Feature extraction

In this study, data collected over a period of one month, encompassing 55,466 data points, was analysed. This dataset, referred to as dataset D1, included the sensed parameters. In dataset D1, the feature vector consists of PM2.5, PM10, CO2, and temperature readings. Binary labels 0 and 1 denote "Not-Cooking" and "Cooking" activities, respectively. Fig. 6 shows an example of an entry in D1.

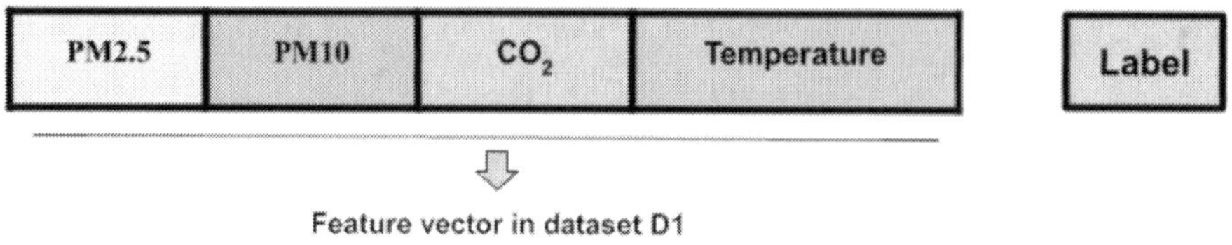

Fig. 6. Entry in dataset D1

3.2.2 Training

From dataset D1, training and testing data were split with 80% samples in the training set and 20% samples in the testing set. With the help of Dataset D1 and corresponding binary labels, an ML model tasked with binary classification was trained. Six different ML models were employed to find the best-performing model: 1. Support Vector Classifier (SVC) 2. Logistic Regression (LR) 3. eXtreme Gradient Boosting (XGB) 4. Random Forest (RF) 5. Naive Bayes (NB) 6. K-Nearest Neighbour (KNN). The machine learning models consider each data point individually for prediction. To improve the predictions by considering the trend of values, we use DL models and pass the data points in sequences of length 100 for training and testing. Four DL models were employed to find the best-performing model: 1. Multi-Layer Perceptron (MLP) 2. Recurrent Neural Network (RNN) 3. Long Short-Term Memory (LSTM) 4. Gated Recurrent Unit (GRU).

3.3 Classification for Day-of-the-Week Assignment

3.3.1 Feature extraction

We analysed data collected over the four months from both the target location and a reference node placed outside the cafeteria in an open environment. This reference node measured PM2.5, PM10, and temperature, providing a standard for comparison. Dataset D2 was created for this task.

For each day, the hourly averages of CO_2, PM2.5, PM10, and Temperature were calculated for the cafeteria kitchen nodes. These 96 (4x24) values comprise the feature vector in Dataset D2. The hourly averages of PM2.5, PM10, and Temperature for the outside node were subtracted from the corresponding feature vector values. This compensates for the environmental effect on the data readings. Numeric labels 0 to 6 were assigned to the days of the week, corresponding to Monday through Sunday. Fig. 7 shows an example of an entry in D2.

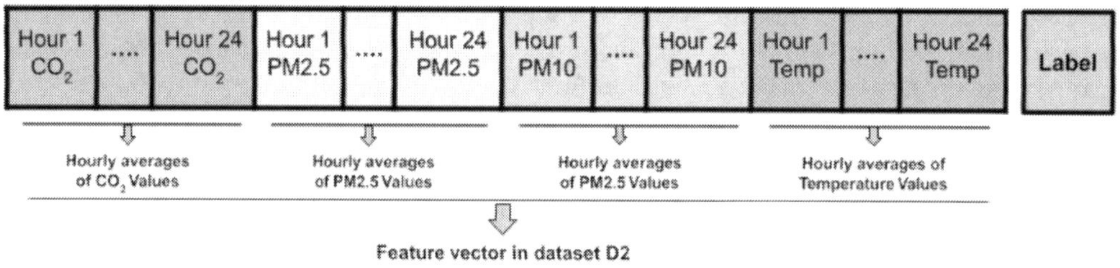

Fig. 7. Entry in dataset D2

3.3.2 Training

With the help of Dataset D2 and corresponding labels, an ML model tasked with multi-class classification was trained. Four different ML models were employed to find the best-performing model based on their ability to perform well with relatively small datasets: 1. Random Forest (RF) 2. eXtreme Gradient Boosting (XGB) 3. Logistic Regression (LR) 4. Decision Tree (DT).

4 Results

This section contains the results for the binary classification task and the day-of-the-week assignment tasks. Four performance metrics have been calculated for all the models to find the best-performing model: 1. Accuracy 2. Precision 3. Recall 4. F1-Score. Comprehensive results and their analysis are covered in this section.

4.1 Binary classification of cooking activity

This subsection contains results from various machine learning and deep learning models trained and evaluated for the task of binary classification to categorise data points into "Cooking" and "Not-Cooking" activities.

Table 2. Performance metrics for machine learning models (in percentages)

	XGB	KNN	RF	LR	SVC	NB
Accuracy	**92.59**	92.54	90.83	90.44	90.03	87.86
Precision	88.74	89.54	84.80	90.48	87.49	78.01
Recall	93.66	92.46	94.37	85.48	88.01	90.84
F1-Score	**91.14**	90.98	89.33	87.91	87.75	83.94

Table 2 shows the performance metrics of the machine learning models. The best-performing machine learning model in terms of accuracy is the XGB classifier. Other machine learning models perform well and are not far behind in metrics. Slightly lower precision across all models may be due to class imbalance, as there are more data points for the "Not-Cooking" class than the "Cooking" class. High recall, in the case of XGB, KNN, and RF, indicates that the models effectively identify most of the true positive instances.

Table 3. Performance metrics for deep learning models (in percentages)

	GRU	LSTM	RNN	MLP
Accuracy	**95.16**	94.73	94.18	90.00
Precision	**92.98**	92.77	90.59	81.29
Recall	**95.07**	94.05	95.35	92.44
F1-Score	**94.01**	93.41	92.91	86.51

Table 3 shows the performance metrics of the deep learning models. The best-performing deep learning model is the Gated Recurrent Unit (GRU). As expected, there is an increase in accuracy from machine learning to deep learning models. Other deep learning models, such as LSTM and RNN, perform well for the task. These models effectively manage sequential data by retaining and utilising contextual information from previous inputs. Our analysis shows that deep learning models outperform traditional machine learning models in terms of precision, suggesting a more effective management of class imbalance in our dataset. Moreover, the F1-score, which is crucial for evaluating model performance in the context of imbalanced datasets, demonstrates superior performance in deep learning models as compared to the machine learning models.

4.2 Classification for day-of-the-week assignment

This subsection contains results from various machine learning models trained and evaluated for the classification task of day-of-the-week.

Table 4. Performance metrics for machine learning models (in percentages)

	RF	XGB	LR	DT
Accuracy	**95.00**	85.00	85.00	70.00
Precision	**96.25**	87.50	86.25	72.92
Recall	**95.00**	85.00	85.00	70.00
F1-Score	**94.86**	83.21	85.86	62.57

Table 4 shows the performance metrics of the machine learning models. The best-performing machine learning model is the Random Forest (RF) which performs exceptionally well for the task across all metrics. It is known to perform reasonably well with smaller datasets. Other models, such as XGB and LR, perform reasonably well for the task. The results support our claim that different food items and cooking practices have different identifiable air pollution signatures.

5 Conclusions

Results from the study show that seemingly trivial air pollution data with basic air quality parameters can reveal complex insights, posing a significant threat to privacy. Inexpensive sensors, when calibrated properly and enabled with IoT, can serve for large-scale data collection, pushing data to storage servers ready to be analysed. Our experiments show that granular data with high sensing frequency is not the only requirement for inferences. Even coarse data measures, such as hourly averages for day-of-the-week assignment, can be used for complex inferences with great accuracy. The effectiveness of various machine learning and deep learning models in performing classification tasks underscores that these inferences do not require extensive expertise in the field. This indicates that the data contains inherent patterns that are readily identifiable by these models, even with basic implementation. As a result, the barrier to exploiting such data for privacy-invasive purposes is lowered.

Given the implications associated with air pollution data, it becomes important to carefully monitor the objectives behind its collection and distribution. Allowing access to this data within strictly defined limits is crucial to guarantee ethical use and legitimate purposes. Policymakers and system developers need to collaborate in establishing rules that protect individual privacy while enabling the advantages of environmental monitoring to be realised. With the rapid advancement in machine learning and deep learning, a proactive approach to privacy is needed to ensure that air pollution monitoring or other such data usage does not intentionally or unintentionally breach individual privacy rights.

References

1. World Health Organization, *Household air pollution and health* https://www.who.int/en/news-room/fact-sheets/detail/household-air-pollution-and-health (2023) Accessed: 2024-07-18.
2. M. Mannan and S. Al-Ghamdi, *Indoor air quality in buildings: A comprehensive review on the factors influencing air pollution in residential and commercial structure* International Journal of Environmental Research and Public Health, vol. **18**, p. 3276, (2021).
3. Precedence Research, *Consumer IoT Market Size Is Expanding USD 616.75 Bn By 2032* (2023). Accessed : 2024-07-18.
4. A. P. Singh, V. Jain, S. Chaudhari, F. A. Kraemer, S. Werner, and V. Garg, *Machine learning-based occupancy estimation using multivariate sensor nodes* in IEEE Globecom Workshops (GC Wkshps), pp. 1–6, IEEE, (2018).
5. A. Molina-Markham, P. Shenoy, K. Fu, E. Cecchet, and D. Irwin, *Private memoirs of a smart meter* in Proceedings of the 2nd ACM workshop on embedded sensing systems for energy-efficiency in building, pp. 61–66, (2010).
6. J. Kroger, *Unexpected inferences from sensor data: A hidden privacy threat in the internet of things* in Internet of Things. Information Processing in an Increasingly Connected World (L. Strous and V. G. Cerf, eds.), (Cham), pp. 147–159, Springer International Publishing, (2019).
7. R. L. Rutledge, A. K. Massey, and A. I. Anton, *Privacy impacts of iot devices: A smarttv case study* in IEEE 24th International Requirements Engineering Conference Workshops (REW), pp. 261–270, (2016).
8. Gerina, F., Massa, S.M., Moi, F. *et al. Recognition of cooking activities through air quality sensor data for supporting food journaling* in Hum. Cent. Comput. Inf. Sci. **10**, 27 (2020).
9. Prana Air, *Prana air indoor pm sensor specification,* https://www.pranaair.com/in/air-quality-sensor/indoor-pm-sensor Accessed: 2024-07-20.
10. Prana Air, *Prana air co2 sensor specification,* https://www.pranaair.com/air-quality-sensor/carbon-dioxide-co2-sensor Accessed: 2024-07-20.
11. R. Bose, A. Parmar, H. Narla, and S. Chaudhari, *Comparative evaluation of low-cost co2 sensors for indoor air pollution monitoring* in IEEE 8th World Forum on Internet of Things (WF-IoT), pp. 1–6, (2022).
12. Aeroqual, *Series 500 – Portable Air Quality Monitor,* https://www.aeroqual.com/s-series-portable-air-monitors/series-500-portable-air-pollution-monitor

    Accessed: 2024-07-20.

13. A. Parmar, S. Sara, A. K. Dwivedi, C. R. Reddy, I. Patwardhan, S. D. Bijjam, S. Chaudhari, K. S. Rajan, and K. Vemuri, *Development of end-to-end low-cost IoT system for densely deployed pm monitoring network: An indian case study* in Front. Internet. Things, vol. **3** (2024).

Citywide mobile air quality monitoring using GPS-enabled low-cost IoT sensors

Shreyash Gujar[1*], *Hitesh* Pamireddy[2], *Naga Saiteja* Maradani[2], *Spanddhana* Sara[1], *Sachin* Chaudhari[1], and *K. S.* Rajan[1]

[1]International Institute of Information Technology-Hyderabad (IIIT-H), India
[2]BV Raju Institute of Technology-Narsapur (BVRIT), India

Abstract. Particulate matter (PM) is a critical air pollutant with severe health implications, yet existing stationary monitoring networks often fail to capture its complete spatial and temporal variability in urban environments. This paper presents a novel approach to city-wide air quality monitoring using low-cost sensors mounted on mobile platforms. To validate this method, a six-month field study was conducted in Hyderabad, India, deploying IoT-enabled devices on four college buses. These mobile devices capture PM concentrations, temperature, relative humidity, GPS coordinates, and vehicle speed twice daily across different parts of the city. The primary objective was to demonstrate the necessity of mobile air pollution monitoring to identify PM variability across diverse urban areas. The data revealed significant variations in PM concentrations across different parts of the city and seasons, highlighting the impact of local activities on air quality. The study examines seasonal trends, area-specific variations, and temporal patterns of PM concentrations, identifying pollution hotspots within the city. It shows how important it is to provide up-to-date, location-specific air quality information to people with pollution-related health issues.

Keywords: Air pollution, Dataset, IoT, Low-cost, Mobile sensing, PM, Seasonal, Spatial, Temporal.

1 Introduction

Air pollution has emerged as a critical global concern, posing severe threats to human health and environmental well-being [1]. Particulate Matter (PM), a complex mixture of airborne particles including dust, soot, and aerosols, poses a significant threat to air quality [2]. PM is classified based on size, with PM2.5 (diameter ≤ 2.5 micrometres) and PM10 (diameter ≤ 10 micrometres) being of primary concern due to their ability to penetrate deep into the respiratory system and potentially enter the bloodstream [3]. In Indian urban centres, PM pollution is especially severe, with air quality deteriorating due to emissions from stationary

[*] Corresponding author: shreyash.gujar@research.iiit.ac.in

and mobile sources. Many Indian cities consistently rank among the world's most polluted due to rapid urbanization, industrial emissions, and dense traffic [4].

The Central Pollution Control Board (CPCB) [5] of India has established air quality monitoring stations across various cities. However, the sparse distribution of these stations poses significant challenges in capturing comprehensive air quality profiles. In a large urban area like Hyderabad, a major metropolitan city in southern India covering 650 sq. km with only 14 CPCB monitoring stations, capturing data at 15-minute intervals cannot capture the city's complex pollution patterns. Furthermore, given that pollution levels in Indian cities can vary significantly every 400-500 meters [6], the complex urban environment creates dynamic air quality patterns. Atmospheric conditions, seasonal changes, and the interaction between buildings, traffic, and local weather all contribute to these variations. The existing monitoring systems, however, fail to detect these localized variations or capture the impact of specific events on air quality at a micro-scale.

Previously, to address the limitations of citywide scaling of air pollution monitoring systems, studies [7-11] have explored mobile air quality monitoring, but each study has significant limitations. [7] measured CO and NO2 across Hyderabad using mobile sensors, but their study was temporally limited and did not account for seasonal variations. Similarly, [8] focused solely on PM2.5 hotspots in urban areas, neglecting other pollutants and broader temporal scales. [9] conducted a multi-city mobile data collection in India, but their one-time approach failed to capture long-term trends or seasonal variations. [10] deployed a single node on a motorcycle, but their study was constrained by a short duration (7 days), limited daily collection time (2 hours), and a small area (6 km) while also excluding PM10 measurements. In a more extensive study, [11] used trams for mobile air quality monitoring in Zurich, Switzerland. While this approach allowed for substantial data collection, it mainly focused only on gaseous pollutants, omitting crucial particulate matter measurements.

Despite the limitations of previous studies, which primarily focused on pollutants like PM2.5, CO, NO2, and O3 and often collected data for a limited time or in limited areas, this research introduces a significant advancement by developing and deploying multiple GPS-enabled, low-cost IoT devices on college buses to collect data across various parts of the city and throughout different seasons, with twice-daily measurements.

The main contributions of this study are as follows:

- A low-cost IoT-based GPS-enabled air quality node has been developed and deployed on four college buses for extensive mobile air quality monitoring to demonstrate its efficacy in urban environments.
- Over six consecutive months, a substantial dataset of more than 5 lakh data points was collected across different seasons and areas of Hyderabad, India, at different times of the day.
- PM analysis was performed to analyze seasonal, areawise, and time-based variations to identify PM hotspots across the city.

2 Methodology

This section presents the hardware details of the air quality monitoring node, data collection campaign, preprocessing pipeline, and the subsequent data analysis techniques employed in this study.

 https://doi.org/10.1051/e3sconf/202562001002

2.1 Hardware details

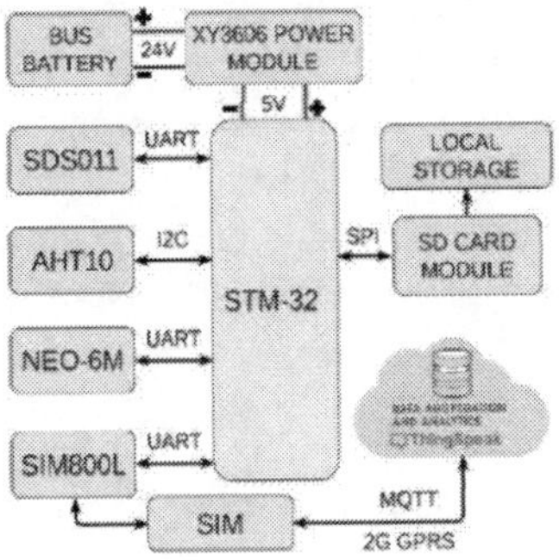

(a) Block architecture

(b) Actual deployed node

Fig. 1. Block architecture and the actual deployed air quality monitoring device.

Fig. 1a illustrates the block architecture of the air quality monitoring node used in this study, while Fig. 1b depicts the actual deployed node. The system utilizes an STM32 microcontroller as the central processing unit. It integrates an SDS011 sensor for measuring particulate matter (PM2.5 and PM10), an AHT10 sensor for temperature and relative humidity (RH) measurements, a NEO 6M GPS module for capturing latitude, longitude, and speed data, and a SIM800L V2 GSM module for transmitting data to ThingSpeak. An SD (Secure Digital) card module is also incorporated for local data storage to prevent data loss. The entire system operates on a 5-volt power supply, which is derived from a 24-volt DC battery in the bus through an XY3606 power module. The nodes are configured to record timestamps, PM2.5, PM10, temperature, humidity, latitude, longitude, and speed at 20-second interval, allowing for the monitoring of rapid changes in air quality. Pre-deployment, these low-cost nodes were calibrated using regression-based models to enhance the accuracy of the data. Refer to [6, 12] for more details regarding the calibration process.

2.2 Data collection campaign

Fig. 2. shows the visual representation of the distinct routes covered by the four buses across Hyderabad, India. The specific routes covered by each bus are as follows:

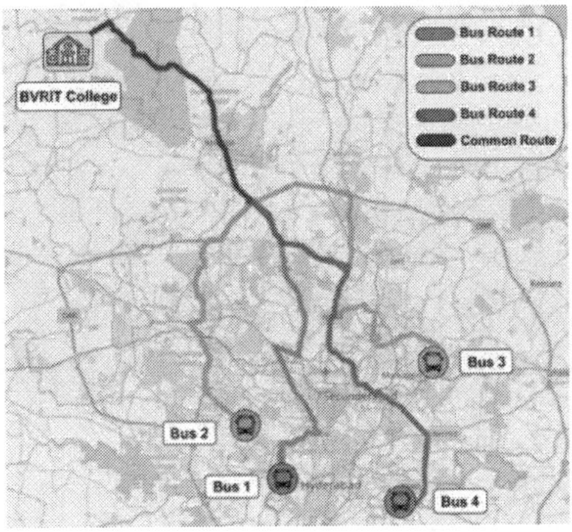

Fig.2. Routes travelled by the four buses. Each route is marked by a different colour for better understanding. (Best viewed in colour).

Bus Route 1: Attapur ⇆ Medhipatnam ⇆ Lakdikapul ⇆ Erragadda ⇆ Balanagar ⇆ IDPL ⇆ Suraram ⇆ Gandimaisamma ⇆ Annaram ⇆ Gummadidala ⇆ Narsapur ⇆ BVRIT.
Bus Route 2: Shaikpet ⇆ Kondapur ⇆ Miyapur ⇆ Bachupally ⇆ Mallampet ⇆ Gandimaisamma (ORR) ⇆ Annaram ⇆ Gummadidala ⇆ Narsapur ⇆ BVRIT.
Bus Route 3: ECIL ⇆ A. S. Rao Nagar ⇆ Suchitra ⇆ Kompally ⇆ Gandimaisamma (ORR) ⇆ Annaram ⇆ Gummadidala ⇆ Narsapur ⇆ BVRIT.
Bus Route 4: L. B. Nagar ⇆ Nagole ⇆ Habsiguda ⇆ Rail Nilayam ⇆ Bowenpally ⇆ Suchitra ⇆ Kompally ⇆ Gandimaisamma ⇆ Annaram ⇆ Gummadidala ⇆ Narsapur ⇆ BVRIT.

Each route begins in a distinct urban area and converges at BVRIT College in the morning. In the evening, the reverse routes are employed, with BVRIT College as the departure point. The buses adhered to a consistent daily schedule throughout the data collection period: morning travel commenced at 06:40 and arrived at its destination by 09:00, while evening travel began at 15:40 and concluded between 18:40 and 19:00, depending on the route. This bi-directional, twice-daily data collection strategy was implemented to capture daily variations in PM levels. To analyze pollution levels across the areas covered by the bus routes, the routes were divided into 25 distinct regions, as outlined in Table 1.

Table 1. Regions considered for this study.

Regions	Start point	End point		Regions	Start point	End point
R1	Attapur	Medhipatnam		R14	LB Nagar	Nagole
R2	Medhipatnam	Lakdikapul		R15	Nagole	Habsiguda
R3	Lakdikapul	Erragadda		R16	Habsiguda	Rail Nilayam
R4	Erragadda	Balanagar & IDPL		R17	Rail Nilayam	Bowenpally
R5	Balanagar & IDPL	Suraram		R18	Bowenpally	Suchitra
R6	Suraram	Gandimaisamma		R19	Suchitra	Kompally
R7	Shaikpet	Kondapur		R20	Kompally	Gandimaisamma (ORR)
R8	Kondapur	Miyapur		R21	Kompally	Gandimaisamma
R9	Miyapur	Bachupally		R22	Gandimaisamma	Annaram
R10	Bachupally	Mallampet		R23	Annaram	Gummadidala
R11	Mallampet	Gandimaisamma (ORR)		R24	Gummadidala	Narsapur
R12	ECIL	AS Rao Nagar		R25	Narsapur	BVRIT
R13	AS Rao Nagar	Suchitra		-	-	-

2.3 Data preprocessing

Fig. 3 presents the data collection and pre-processing pipeline designed to improve the quality of the raw dataset, ensuring the reliability and accuracy of subsequent analyses. The pre-processing workflow initiated with humidity filtering, excluding data points where relative humidity (RH) values surpassed 80%, to reduce potential biases linked to the sensor's limited measurement range, as outlined in the datasheet. Next, PM measurements surpassing 999 $\mu g/m^3$ were removed, consistent with the standard PM measurement range of 0-999 $\mu g/m^3$. Outlier removal was conducted using the Interquartile Range (IQR) method, which calculates the difference between the first quartile (25th percentile) and the third quartile (75th percentile), approximating 1.35 times the standard deviation of the data.

As the sensors were deployed on buses powered by their batteries, instances where drivers inadvertently left the batteries on during stationary periods were identified and the corresponding data was removed to prevent dataset skewing. The initial raw dataset contained 532,083 data points. Following the removal of outliers (6.56% of the original data) and measurements taken during stationary periods (22.22% of the original data), the final processed dataset consisted of 378,949 data points. To address any gaps in the dataset, linear interpolation was applied, ensuring a continuous and consistent data series. Lastly, regression-derived calibration coefficients were applied. This comprehensive pre-processing procedure results in a high-quality processed PM dataset suitable for further analysis.

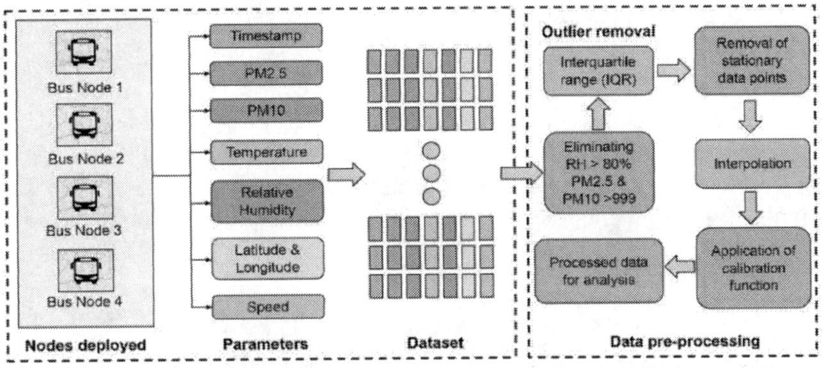

Fig. 3. Data collection and preprocessing pipeline. (Best viewed in colour).

2.4 Data analysis

Following the data pre-processing and application of calibration coefficients, a series of analytical methodologies were employed to comprehensively assess PM levels across various urban areas. Spatial visualization mapping of the processed PM data from each bus-mounted sensor node was conducted and visualized on city maps, facilitating the spatial variability in air quality across the urban landscape. Temporal analysis of the six-month dataset was performed to identify the daily PM level fluctuations, providing insights into urban air quality dynamics over time. Seasonal and daily variation analysis was carried out by generating bar plots that depicted average PM concentrations in distinct city areas, categorized by season (winter, summer, monsoon) and time of day (morning, evening). This analysis revealed significant averaged temporal variations in air quality on both seasonal and daily scales. Temporal and spatial based PM distribution were assessed using box plots for data from all four bus routes, enabling a comparative analysis of PM level distributions across different urban areas. This approach underscored the heterogeneity in air quality experienced along various transit routes.

3 Results

This section presents a comprehensive analysis of PM concentrations, including spatial visualization of PM concentrations across the city, daily variations in PM concentrations, seasonal variations of average PM concentrations, and the temporal and spatial distribution of PM concentrations.

3.1 Spatial visualization of PM concentrations

Fig. 4a and Fig. 4b illustrates the spatial visualization of PM2.5 and PM10 concentrations across the city. The visualization was generated using collected PM data and geospatial methods, overlaying the results on a city map. Both PM2.5 and PM10 values were categorized into distinct ranges and colour-coded accordingly. This visualization allows observing PM variations across different urban areas at different times of day. Analysis of the PM values in conjunction with real-time images reveals that pollutant concentrations fluctuate based on location and road conditions. Areas exhibiting both high vehicular traffic and visible construction sites consistently show PM2.5 levels in the 'moderate' to 'poor' and PM10 levels in the 'very poor' to 'severe' ranges. Notably, satisfactory PM2.5 and moderate PM10 levels were commonly observed even in locations with minimal vehicular presence. This phenomenon may be attributed to ongoing road construction and the resultant dust. These results indicate a strong correlation between vehicular traffic and urban development activities, particularly road construction with elevated PM2.5 and PM10 concentrations throughout the city. The collected dataset through this study lays the groundwork for future research, including the potential development of real-time pollution maps using advanced geospatial techniques.

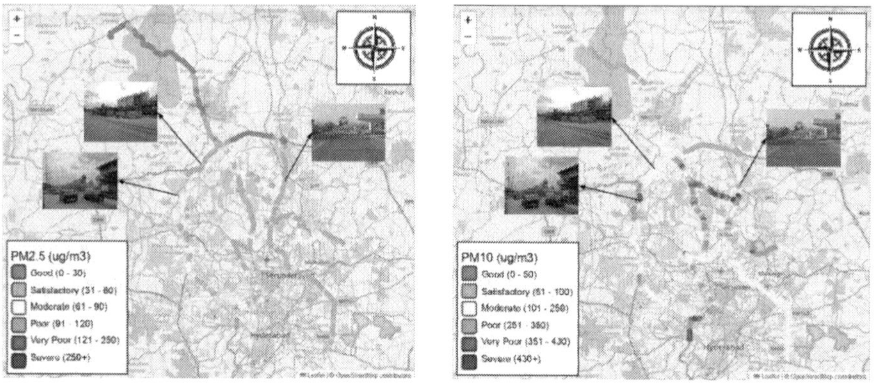

(a) PM2.5 (b) PM10

Fig. 4. The plot shows the spatial visualization of PM2.5 and PM10 concentrations across the city. (Best viewed in colour).

3.2 Daily variations in PM concentrations

Fig. 5 shows a time series plot of daily average PM concentrations for all four bus routes from January to June 2024. This plot provides a clear view of how PM2.5 and PM10 levels change over different months and seasons. In January, which represents winter, PM2.5 levels constantly range between 50 ppm to 150 ppm, while PM10 levels generally show values more than 250 ppm range for all bus routes, with some bus routes (1 and 2) occasionally crossing the 350 ppm range. This pattern is typical for winter pollution in cities. From February to May, during the summer season, PM levels show a gradual improvement. In February and March, PM2.5 levels consistently range between 50 ppm and 100 ppm for all the bus routes, while the PM10 levels for bus routes 1 and 2 fluctuate between the 200 ppm to 300 ppm range and bus routes 3 and 4 stay below the 200 ppm range. As summer continues, especially in April and May, PM2.5 values consistently fall around the 50 ppm range, and PM10 levels consistently fall within the 100 ppm to 200 ppm range. In June, which marks the start of the monsoon season, air quality improves significantly. On rainy days, bus routes 2 and 4 show PM2.5 and PM10 levels below the 100 ppm range. This highlights how rainfall can significantly reduce PM10 levels. Overall, this time series plot effectively demonstrates how PM levels vary with the seasons and weather conditions, showing clear seasonal trends and how precipitation can quickly improve air quality. A consistent trend was observed between PM2.5 and PM10 values, so for the further analysis, only PM10-based results will be presented for brevity.

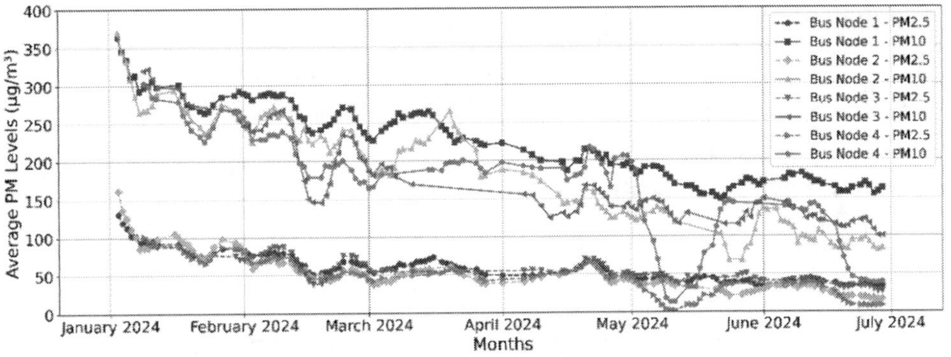

Fig. 5. The time series plot displays the daily average PM2.5 and PM10 levels across six months, which also shows the effect of seasons on PM levels. (Best viewed in colour).

3.3 Seasonal variation of average PM10 concentrations

Fig. 6a and Fig. 6b illustrate bar plots depicting the average PM10 levels across various areas in Hyderabad, covering three distinct seasons: winter, summer, and monsoon. For this analysis, January 2024 represents winter, February to May 2024 represents summer, and June 2024 represents the monsoon season. The data is further segmented into morning and evening measurements. The PM10 levels display consistent variation across all areas, with a clear seasonal pattern emerging. Winter consistently records the highest PM10 levels, followed by

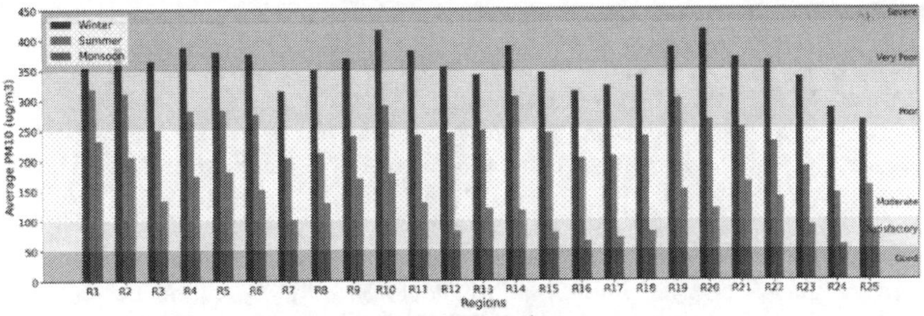

(a) Morning plot

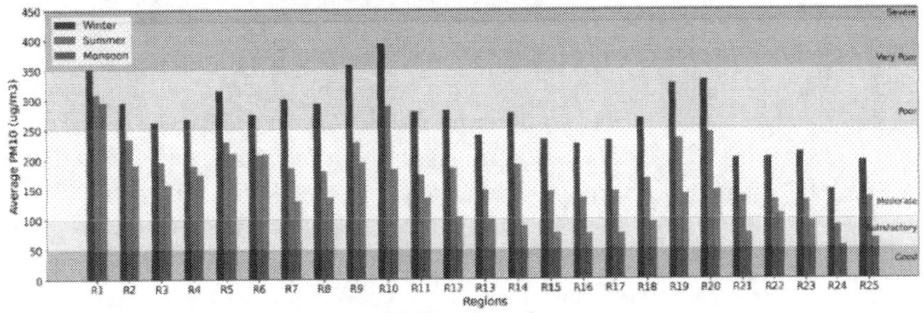

(b) Evening plot

Fig. 6. The bar plot displays the average PM10 concentrations across the regions considered for this study during different seasons. Out of the three bars, the first bar shows winter followed by summer and monsoon, clearly showing the effect of location and seasonality on PM10 levels. The plot's background colours are based on PM10 ranges. (Best viewed in colour).

summer, with the lowest levels observed during the monsoon season. This trend is likely influenced by seasonal variations in meteorological conditions. Notably, during both winter and summer, PM10 levels frequently exceed the 'poor' air quality threshold, particularly during morning hours. Morning measurements generally exhibit higher values compared to evening readings across these seasons, likely due to factors such as morning rush hour traffic, temperature inversions that trap pollutants near the ground, lower wind speeds that limit pollutant dispersion, and increased humidity that causes fine particles to settle closer to the surface. These results highlight the strong effect that both the season and time of day have on PM10 levels, showing how city activities, weather, and air quality are closely connected.

3.4 Temporal-based distribution of PM10 concentrations

Fig. 7a and Fig. 7b show the distribution of PM10 concentrations across all four bus routes during the morning and evening periods, respectively. These plots, based on the entire dataset, reveal clear patterns of PM10 variations across different urban routes covered by the buses and times of day. It was observed that there is much higher variability in PM10 levels during the morning hours compared to the evening, which aligns with the atmospheric conditions and daily trends discussed earlier.

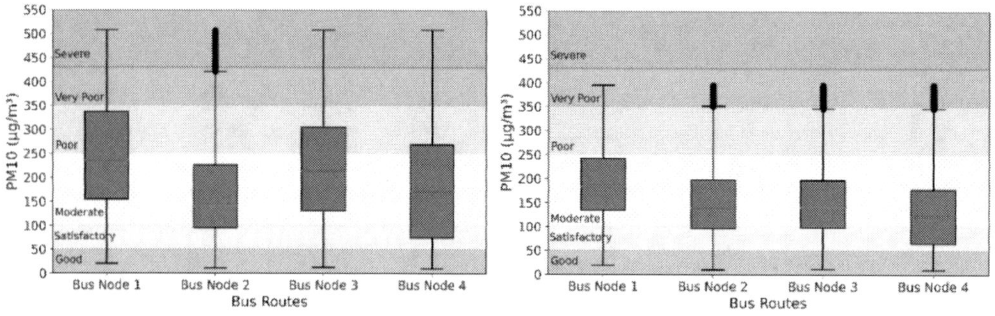

(a) Morning (b) Evening

Fig.7. The box plots display the overall distribution of PM10 levels for all four bus nodes during morning and evening, showing how different routes and times of day can affect the PM10 concentrations. The plot's background colours are based on PM10 ranges. (Best viewed in colour).

Node 1, which monitors the city centre route, exhibits PM10 concentrations predominantly within the moderate to poor range. This can be attributed to ongoing construction activities and the route's central location, which experiences high vehicular traffic throughout the day. Node 2, which follows a route slightly outside the city centre, consistently shows PM10 levels in the moderate range during both morning and evening times. Although this route has steady traffic, its location on the outskirts might help keep the PM10 levels relatively stable. The route monitored by Node 3 has higher PM10 levels, mostly in the moderate range but often reaching the poor category in the morning. Despite having less traffic, the ongoing construction of bridges and roads seems to be a major cause of pollution. Node 4, which goes through areas with different levels of traffic, shows PM10 levels slightly lower but like those recorded by Node 1. This pattern highlights the relationship between traffic, urban development, and air quality in various parts of the city at different times of day. The results highlight the considerable variation in pollution levels across the city and emphasize the importance of accounting for both traffic and local sources, such as construction when assessing changes in urban air quality.

3.5 Spatial-based distribution of PM10 concentrations

Fig. 8a through Fig. 8d present boxplots for each bus route to further explain these patterns, illustrating region-specific PM10 level variations. Bus node 1 predominantly recorded PM10 levels in the moderate to poor range, with region R1 extending into the 'very poor' range, identifying it as a significant PM10 hotspot along with regions R2, R5, R6, and R22, which exhibited slightly lower but similar range values, also indicating elevated pollution levels. Regions covered by bus node 2 showed comparatively lower PM10 levels than those of bus node 1, with most regions falling within the moderate range. However, regions R9 and R10 emerged as localized PM10 hotspots, with values in the 'poor' range. Bus node 3 covered regions with PM10 levels similar to bus node 2, primarily in the moderate range. Notable exceptions were regions R19 and R20, which almost surpass the 'poor' range and identify them as potential PM10 hotspots. Bus node 4, covering the most extensive regions, showed moderate PM10 levels across most regions, with regions R14, R19, and R20 reaching the 'poor' range and thus classified as PM10 hotspots. These observations highlight the spatial heterogeneity of air pollution within the regions considered for this study. The distinct PM10 concentration profiles exhibited by each region underscore the influence of local environmental factors, traffic patterns, and urban characteristics on air quality.

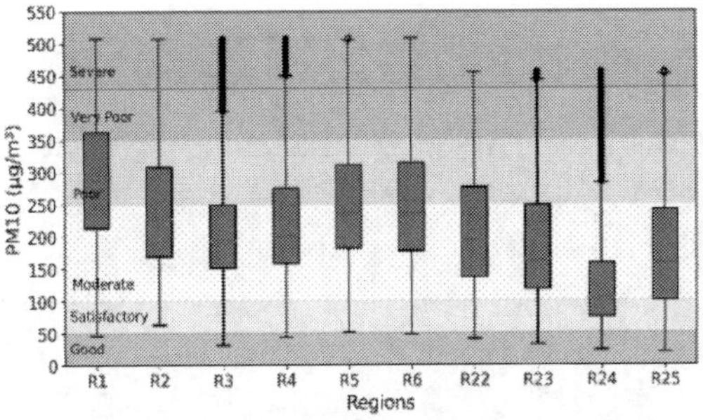

(a) Bus node 1

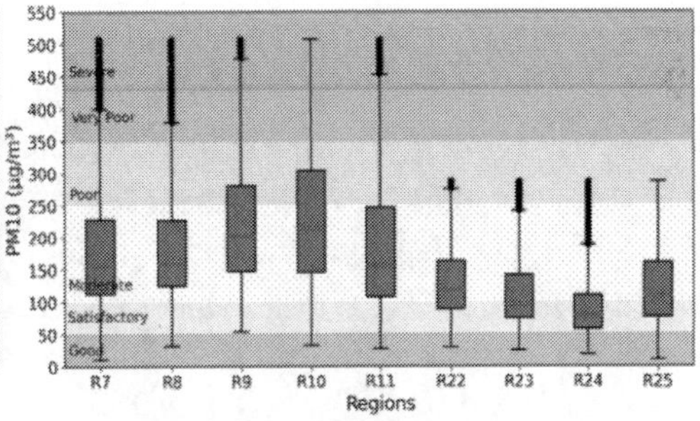

(b) Bus node 2

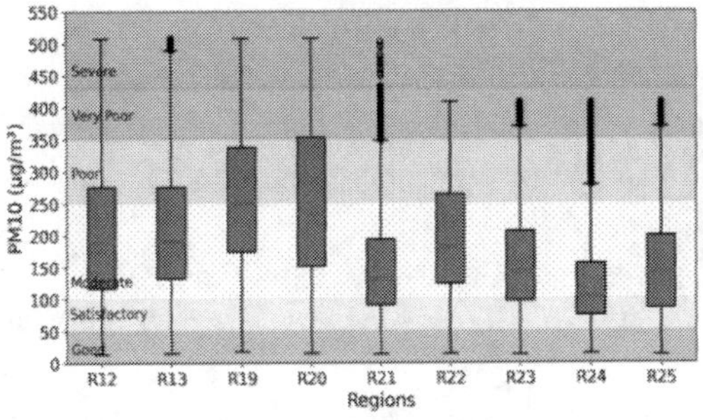

(c) Bus node 3

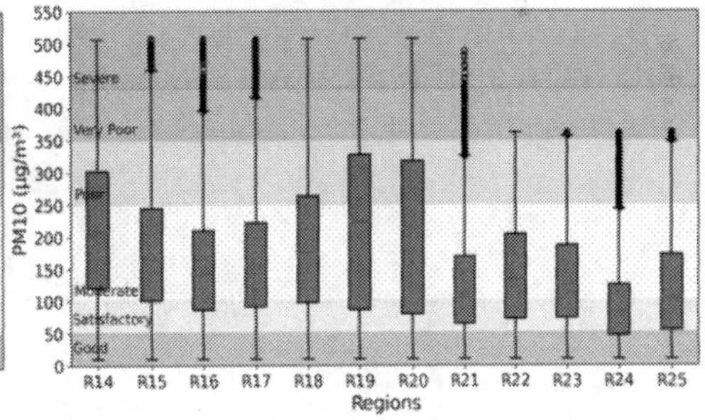

(d) Bus node 4

Fig. 8. The box plots display the distribution of PM10 data across different regions considered for this study, showing how different regions can affect the PM10 concentrations. The plot's background colours are based on PM10 ranges. (Best viewed in colour).

4 Conclusion

The study highlights the crucial role of mobile air quality monitoring in understanding urban pollution patterns. By deploying GPS-equipped monitoring nodes on buses traversing various routes, the study provided a detailed view of PM10 levels city-wide and throughout various times of the day. This approach not only revealed significant variations in air quality depending on traffic density, local construction activities, and seasonal changes but also enabled the detection of PM10 hotspots. For instance, PM10 levels were consistently higher in the morning reaching above 350 ppm followed by 250 ppm in the evening due to rush hour traffic and were influenced by construction work and varying traffic patterns across different routes. These findings underscore the importance of mobile monitoring in capturing the dynamic nature of air pollution. Unlike stationary monitors, mobile nodes can track air quality changes across diverse urban environments and temporal scales, offering a comprehensive and nuanced understanding of how air quality fluctuates. The study's results emphasize how mobile monitoring can reveal patterns and trends that might be missed by traditional stationary networks, particularly highlighting the impact of specific urban factors, meteorological conditions, and the identification of PM10 hotspots on air quality.

Shreyash Gujar thanks IHub-Data, IIIT Hyderabad, for the Research Fellowship. This work was supported by the ECE department of BV Raju Institute of Technology, Narsapur, Telangana, India, which facilitated the deployment of sensor nodes on their college buses.

References

1. World Health Organization. (2021). "Air Pollution." WHO Global Health Observatory.
2. U.S. Environmental Protection Agency.(2021). "Particulate Matter Basics." EPA.gov.
3. Kim, K. H., Kabir, E., & Kabir, S. (2015). "A review on the human health impact of airborne particulate matter." Environment International, 74, 136-143.
4. IQAir. (2021). "World Air Quality Report. Region & City PM2.5 Ranking." IQAir.com.
5. Central Pollution Control Board India. 2015. Report on AQI Control of Urban Pollution Series. http://app.cpcbccr.com/ccr_docs/FINAL-REPORT_AQI_.pdf.
6. A. Parmar et al., "Development of end-to-end low-cost IoT system for densely deployed PM monitoring network: an Indian case study," Frontiers in the Internet of Things, vol. 3, 2024.
7. S. Sara et al., "The Application of Mobile Sensing to Detect CO and NO2 Emission Spikes in Polluted Cities," IEEE Access, vol. 11, pp. 79 624–79 635, 2023.
8. S. Sara and S. Gujar et al., "Protocol for hunting PM2.5 emission hot spots in cities," in Proceedings of the 1st International Workshop on Advances in Environmental Sensing Systems for Smart Cities, ser. EnvSys '23. Association for Computing Machinery, 2023, p. 37–42.
9. S. Deb et.al, "Spatial Factor Analysis of Mobile IoT Data: A Case Study on PM Across India," *IGARSS 2022 - 2022 IEEE International Geoscience and Remote Sensing Symposium*, Kuala Lumpur, Malaysia, 2022, pp. 1272-1275.
10. Y. -L. Chiang et al., "Mobile Measurement of Particulate Matter Concentrations on Urban Streets: System Development and Field Verification," in IEEE Access, vol. 8, pp. 197617-197629, 2020.
11. Li, Jason et al., Jan. (2021). Sensing the Air We Breathe — The OpenSense Zurich Dataset. Proceedings of the AAAI Conference on Artificial Intelligence. 26. 10.1609/aaai.v26i1.8163.
12. I. Patwardhan et al., "Comparative Evaluation of New Low-Cost Particulate Matter Sensors," in 8th International Conference on Future Internet of Things and Cloud (FiCloud), 2021, pp. 192–197.

E3S Web of Conferences **620**, 02001 (2025) https://doi.org/10.1051/e3sconf/202562002001
ICEPP 2024

Adhesion of dust particles and its effects of the material of surface

Ghani Albaali* *and Tala* Al-Masri

Princess Sumaya University for Technology, Amman, Jordan

Abstract. The effect of angle inclination and the material of the surface on the amount of dust collected have been investigated. This parameter was found to vary with the angle of inclination for the same material and with the material for the same angle. This variation bears no correspondence to the roughness of the surface as inferred from the respective values of the coefficient of friction. This selectivity could lend itself to selective deposition and the separation of particles of different sizes. The composition of dust particles and fibers accumulated on glass sheets, 20x20 cm^2, placed horizontally and tilted at different angles (50°, 70°, 80°, 90°, and 120°) was examined. The amount of slipping off the dust from the surface of glass sheets with different tilt angles have been measured.

1 Introduction

Dust is almost frequent in many different countries, and it is one of the eminent characteristics of summer, specifically in the MENA region where large amounts of dust are raised by air. In general, for the dust to be raised and remain in the air, three conditions must be applied, which are dry land surface, moderate wind speed to disturb the dust, and unstable air if thorough vertical motion is to occur. The turbulence caused by the wind in stable conditions. In stable conditions, turbulence caused by the wind is checked and the dust rises few meters.

Another type of common dust storm occurs with strong winds and when the conditions are unstable, some dust storms may last for many hours, and sometimes they extend to the night, but they are less severe, due to the decrease in lapse rate. Most countries in the Arabian Gulf countries are surrounded by deserts which are covered by calcareous large sediments, such as gravel, sand, and clay. These countries are frequently under strong air winds due to the lack of vegetation covering their land. As evidence by the results achieved from a lot of studies which were carried by other researchers related to the dust phenomenon, it is seen that the dust problem is not only a local one but has a regional magnitude. Therefore, Meteorological Departments, Scientific Research centers, and Environmental Protection Departments in the neighbouring countries cooperate and coordinate their activities to investigate the dust phenomenon related to its source, occurrence, transportation, and sedimentation. It is felt that much effort is needed on local and regional bases for the fixation

* Corresponding author: A.Albaa@psut.edu.jo

of dunes by such methods as spraying petroleum derivatives and by utilizing windbreakers such as green belts and similar artificial means.

There are several studies that investigated dust particle adhesion to surfaces include Liu et al. [1] for the effect of dry dust removal for filter material, Plumier et al. [2] for the content of dust and the characteristics of adhesion of corn samples, Tong et al. [3] for adhesion and desorption, Said et al. [4] for the influence of dust accumulation of environmental factors on photo-voltaic (PV) systems and the mitigation strategies of dust-accumulation, Tan et al. [5] for an air-conditioned environment and Said et al. [6] for the influence of dust fouling on the performance of PV module. Other various physical and chemical processes are influenced by particle size or by the size distribution of the material involved. The effect of deposition of dust on photo-voltaic solar panels in desert environment is also presented [7]. The adhesion to surfaces of powder and dust has been extensively dealt with by Zimon [8], who reviewed and discussed the measurement of the forces involved. In HELIO It has observed that the dust deposited on the surface slides off in successive layers at successive angles of tilt. They also observed that the dust samples collected sequentially differed in their size distribution profiles. Some of these effects have been utilized to advantage in the procedure for the measurement of particle size [9 and 10]. In such studies, they include such phenomena as the variation with a particle size of rate of setting in various fluids, variation of electrical resistance and variation of osmotic pressure. Size distribution for airborne particles are functions of the size distribution profiles of the materials being processed, heir densities and shapes (symmetric particles, fibers, etc.), and the aerodynamics of the space in which they are present.

2 Data and methodology

2.1 Data

Dust was allowed to accumulate on the surface of different materials during a specified period. Dust accumulating on the surface (both horizontal and inclined) at an angle alpha was collected and the number of particles, and fibers were measured. The light source for this system is He-Ne laser, beam diameter 5mm, optical power 0.8 mw. Wavelength 632.8 um. The computer analysis system worked under computer control.

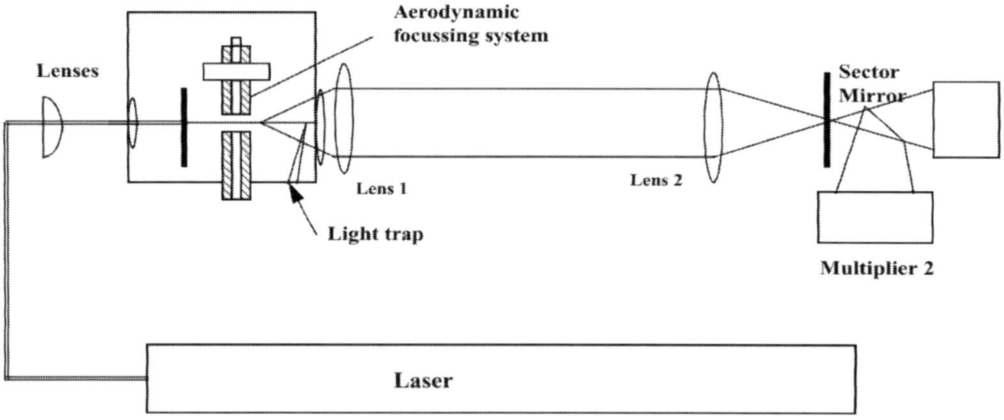

Fig. 1. Laser optical particle counter with fiber recognition.

The highly sensitive aerosols measuring system shown in Figure 1 above consists mainly from the dust photometer for continuous monitoring of dust and measuring its density in mg/m3, finally, the laser optical counter of particles and fibers. These two equipment working under computer control via microprocessor control unit. Figure 2, shows the direction of rotation of the plate for dust collection, and the change in the amount of sliding off dust with angle of inclination for plates A and B (weight of dust sliding off with angle variations).

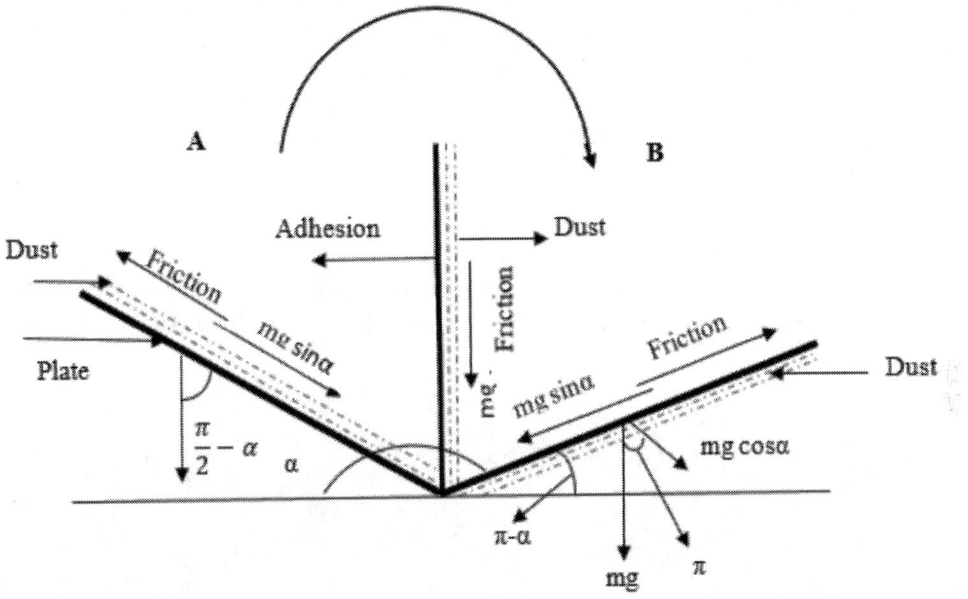

Fig. 2. Sketch showing the direction of rotation of the plate for dust collection.

2.2 Methodology

Two identical sheets of glass with dimensions of 20x20 cm2 were placed horizontally (flat) on the rooftop of the building of Solar Energy Research Center (SERC) and exposed to the atmosphere for thirty days, where the dust is accumulated by the gravity forces. The procedure steps were based on tilting the sheet of glass at an angle α, collecting the dust and weighting the sliding off dust. The delivered software allows running the dust lab in a quite universal way. Light microscope for dust observations is also used and other tools for studying absorption of light and the diffraction on contaminated mirrors or glasses. The software is consist of three program steps for operating the system, which is:

- Operating software
- Measuring software
- Analysis software for stored results and data.

3 Result

From the measurement, the fiber to particle ratio and its average value for suspended dust throughout dust collection were calculated. Table 1 shows the weight per unit area in gm/m^2 deposited on the materials, for sheets placed in horizontal position and sheet placed inclined at an angle 30°. The sheets 20x20 cm were placed facing south and at a height of 19.11 m, during the period (9 am of 25th of July to 2.30 pm of 9th of August).

Table 1. The angle of slide α (M/M) and α (G/M) and weight of dust collected at α = 30° for varions materials M.

Material	Angle (M/M)	Angle (G/M)	Weight of dust collected (mg/m^2) at 30°
Transport glass	13	13	471.1
Transport glass (horizontal)	-	-	773.3
Galvanized steel	14	16	388.4
Shaded glass	20	15	1635.6
Shaded mirror	21	25	396.7
Copper	22	21	444.4
Transparent plastic	24	20	191.1
Aluminum	25	19	880.0
Stainless steel	26	17	404.4

The angle at which slabs of glass 5x5 cm just begin to slide off surfaces of the different materials α (G/M) were measured, so where the angles of slide of blocks of the different materials on surfaces of the same material α (M/M). Figure 3 shows the particle size distributions for dust collected on sheets inclined at 30o for the different surface materials.

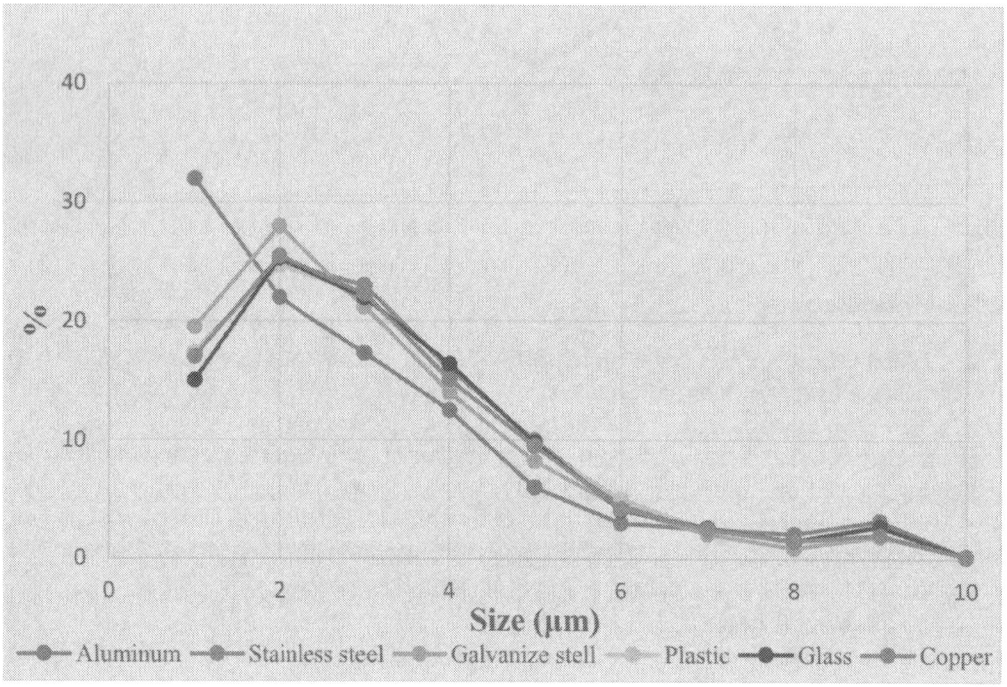

Fig. 3. Particle size distribution of dust collected on sheets inclined at 30°.

Although data deposited on the different surfaces simultaneously, one observes that the mass collected at α =30° was not constant, and vary with the surface material used. The variation bears no relation to α (M/M) or α (G/M), showing that the effect is not friction caused. There is also no similarity in trend with the F/P ratio, which in turns differs for the horizontal and inclined positions. It thus evident that the only factor with influences the size distribution profile is the composition of the material of the surface. The particles of dust contaminate and held on surfaces by the effect of gravity pull, as well as, by the electrostatic attraction, where the layers of dust successively pile up on the top of the adhesion held layer. When the surface is tilted by an angle α, the layers of the dust start to slide off, one over other, $mg \sin \alpha$ for each layer exceeds the friction of interlayer.

The change in the weight of sliding of dust with inclination angle of 50°, 70°, 80°, 90° and 120° was found 33, 101, 27, 26 and 4 for plate A and 20, 131, 23, 8, and 3 for plate B respectively. From the data above, mg sin α and mg cos α (for Figure 2 below) can be calculated. It shows the variation of the friction force mg sinα, parting force mg cosα and the weight of sliding of dust with inclination angle (α) for plates A and B, while Figure 4 shows the change in the weight of sliding off dust with inclination angle for the same plates (A and B).

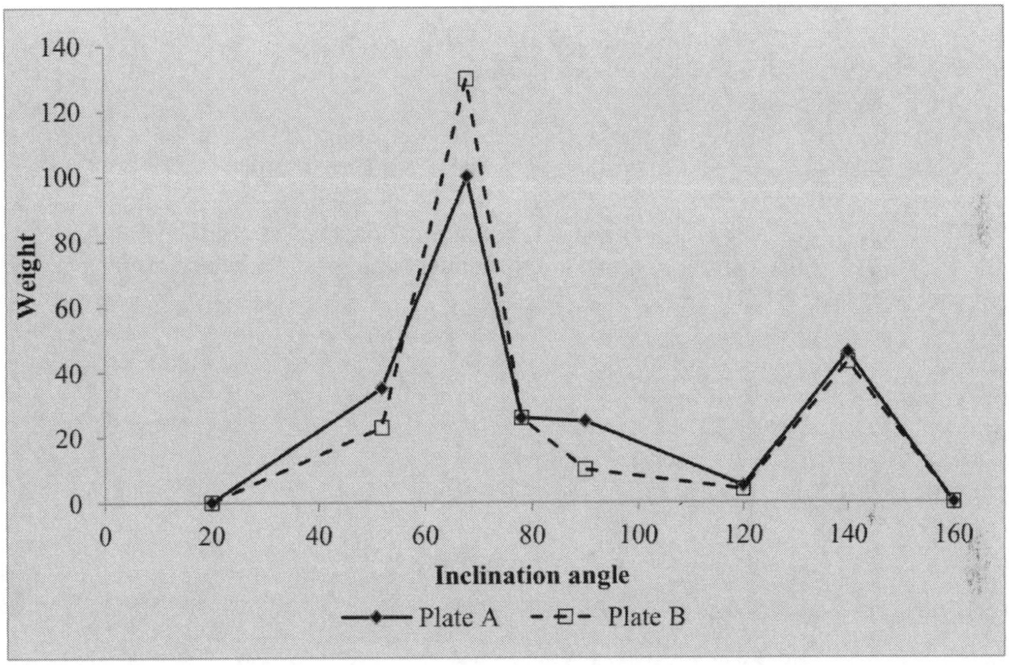

Fig. 4. Variation of the amount of sliding off dust with inclination angle of plates A and B.

From the first glimpse, it is obvious that the size distribution profile of collected dust at α = 50°, 70°, and 80° (i.e. different angle of inclination) are vastly different. If these profiles were compared with the inclination angle α, one conclude that the different profiles of layers are prevented from sliding of different friction forces. This means that the friction forces have been dependent on the profiles of size distribution of stationary and sliding layers.

4. Conclusion

The composition of dust particles and fibers accumulated on glass sheets, 20x20 cm^2, placed horizontally and tilted at different angles (50°, 70°, 80°, 90°, and 120°) was examined. The amount of slipping off dust from the glass sheets surface with different tilt angles have been measured.

It is quite important to do such measurements to see the influence of deposition of dust on the solar systems such as solar cells, solar collectors, solar concentrators or the amount of dust accumulating on electronic instruments. This is in addition to study all related and affected parameters on its performance such as size distribution profile and weight profile, and measure the reduction in the intensity of light as a result of the accumulation of this type of dust. Further studies involving various surfaces and various coating would be needed.

References

1. R. Liu, G. Zhou, C. Wang, W. Jiang, X. Wei, J. of Hazardous Mat., 397, 122734 (2020).
2. B. Plumier, Y. Zhao, M. Casada, R. Maghirang, K. Ambrose, K. (2020). Trans. of the ASABE, **63**(2), 495-499 (2020).
3. L. Tong, X. Chen, Y. Zhang, S. Yin, C. Liu, L. Wang, Y. Ding, Powder Tech., 354, 760-764 (2019).
4. S.A. Said, G. Hassan, H.M. Walwil, N. Al-Aqeeli, *Renewable and Sustainable En. Rev.*, **82**, 743-760 (2018).
5. S.A. Said, H.M. Walwi, Solar En. **107**, 328-337 (2014).
6. C.L.C. Tan, S. Gao, B.S. Wee, A. Asa-Awuku, B.J.R. Thio, Aerosol Sci. and Tech., **48**(5), 541-551 (2014).
7. M. Saidan, A.G. Albaali, E. Alasis, J. Kaldellis, Renewable En. **92**, 499-505 (2016).
8. A.D. Zimon, A. D. (1982). Adhesion of dust particles (Springer, Second Edition, Consultants Bureau, New York, 2018).
9. T. Allen, Particle size measurement (Third Edition. Chapman and Hall, 1981).
10. R. Irani, C. Callis, Particle size measurement interpretation and application (John Wiley and Sons Inc., 1963).

 https://doi.org/10.1051/e3sconf/202562002002

Removal of trace ammonia by ionic liquids

Chih-Han Hsu[1*], *Wei-Chih* Lee[2], *Jiun-Jen* Chen[3], *Yu-hao* Kang[4]

Green Energy and Environment Research Laboratories, Industrial Technology Research Institute, Hsinchu 31057, Taiwan

Abstract. Ammonia (NH_3), a critical process gas in semiconductor production, is the second largest air pollution from semiconductor factories as well as one of PM2.5 precursors. NH_3 emission standard will be regulated more strictly from current 2.5 ppm to 0.8 ppm as revealed by the 2025 semiconductor air pollution standard draft. NH_3 emission concentration was controlled using wet scrubber to achieve the regulations by adjusting pH value and increasing the water-recycling rate. However, it resulted in increased water consumption and the NH_3 exhaust concentration still above emission standards (1 ppm). The purpose of this project was to develop ionic liquids which have high absorption ability to remove low NH_3 concentration. The results show that IL-1 and IL-5 were both good for NH_3 absorption, which the absorption capacity was greater than water. Among them, the liquid level of IL-1 will increase after the NH_3 absorption test, which indicates that it has good water absorption capacity. It has the dual functions of dehumidification and ammonia removal. Furthermore, IL-5 has the best NH_3 absorption ability which compared at same concentration can maintain the NH_3 outlet concentration at 0 ppm for more than 351 minutes.

1 Background

Among the air pollutants emitted by semiconductor factories over the years, NH_3 (ammonia) was the second largest amount of air pollution emissions from semiconductor factories (Fig. 1). It is also the main precursor of $PM_{2.5}$ (15-35% originates from NH_3). The exhaust facility of semiconductor factories were existed the problem of low concentration NH_3 emission due to poor recycled water quality and insufficient efficiency of water washing treatment equipment. The central scrubber of the terminal air pollution treatment equipment currently adopts the method of adjusting the washing water volume and pH value, increasing preventive maintenance times to improve washing efficiency, but trace amounts of NH_3 are still directly exhausted into the atmosphere(the emission concentration is higher than 1~2 ppm). By 2025, the regulations for semiconductor air pollution emission standards will be significantly tightened. NH_3 emission standard needs to be reduced from 2.5 ppm to 0.8 ppm.

Currently, recycled water is utilized as the working fluid for the wet local scrubber (L/S) during waste gas treatment. To improve the removal efficiency of acidic and alkaline gases, acid and alkali chemicals are added. However, the main source of NH_3 is due to two factors. Firstly, wet L/S is not efficient in removing NH_3. Secondly, during wet scrubbing processes,

* Corresponding author: juliehsu@itri.org.tw

NH_3 is easily stripped into the pipeline by the recycled water. The objective of this research is to develop liquid materials that can enhance NH_3 absorption efficiency. The plan is to directly load them in a wet scrubber or install them separately in the backend (Fig. 2). To address the issue of low-concentration escape and minimize the use of recycled water in wastewater treatment, we provide semiconductor factories worldwide with a technology that removes trace NH_3 through high air volume and low energy consumption.

Ionic liquids (IL) have the advantages of extremely low vapor pressure, good thermal stability, good gas solubility, and structural designability. In recent years, they have been widely used in ammonia separation and recovery technology [2-3]. This study aims to develop ionic liquids that can absorb low concentrations of NH_3. Zeng et al. stated that introducing weakly acidic protons through cations can provide polar functional groups (such as OH groups) to form intermolecular hydrogen bonds with NH_3 [3]; Osborne et al. introduce anions into metal ions to provide metal coordination sites and NH_3 to form Lewis acid-base forces [4]. The ionic liquid absorption mechanism is shown in Fig. 3. The cation and anion structures are designed to have multiple active sites and exert a synergistic effect. This greatly increases the interaction force with NH_3 and improves NH_3 absorption ability.

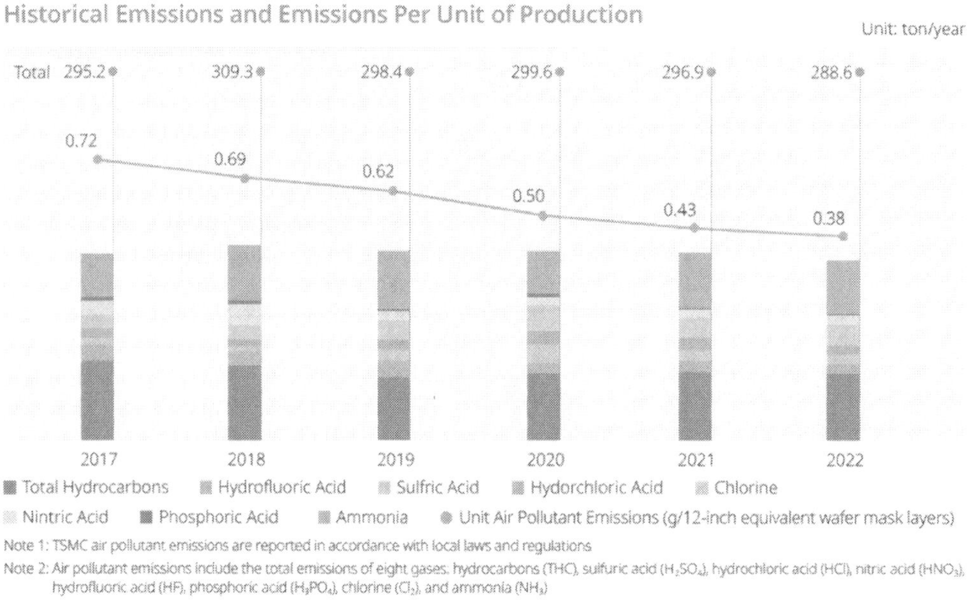

Fig. 1. Historical Emissions and Emissions Per Unit of Production [1]

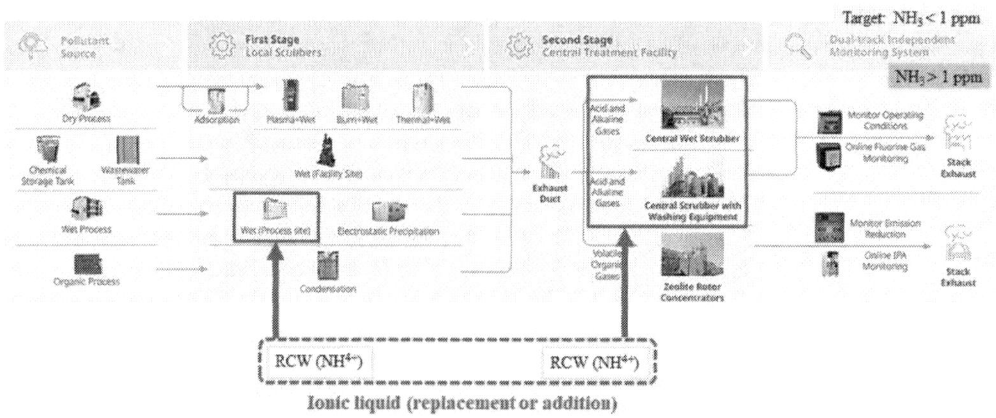

Fig. 2. Ionic liquids are applied to wet scrubbers in semiconductor factories.[1]

https://doi.org/10.1051/e3sconf/202562002002

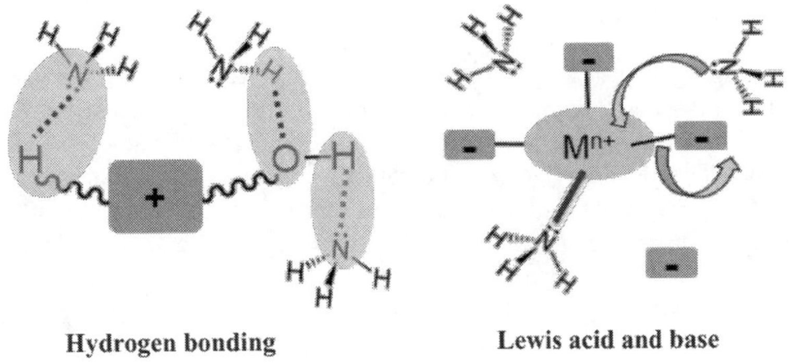

Fig. 3. NH_3 absorption mechanism of ionic liquids

2 Experimental

2.1 Material

NH_3 (10%) was purchased from New Gas Co., LTD. Chemical reagents of analytical grade were purchased from Uni-Onward Corp. The gas flow rate in the system was controlled using mass flow controller(Model: 5850E). The measurement of NH_3 concentration was detected by NH_3 gas monitor (POLI, range: 0-500 ppm, resolutions: 1 ppm).

2.2 Synthesis of ionic liquid

Ionic liquids were synthesized in the following steps (as shown in Fig. 4)

1. Follow the steps of heating reaction, filtering impurities, purification, extraction and vacuum concentration to synthesize the ionic liquid stock solution.
2. Ionic liquid stock solution were send to NMR and ESI-MS for structural analysis.
3. Prepare different concentrations of ionic liquid, and confirm the volumetric water content according to Karl Fischer
4. Adjust different synthesis parameters and repeat steps 1~3 above.

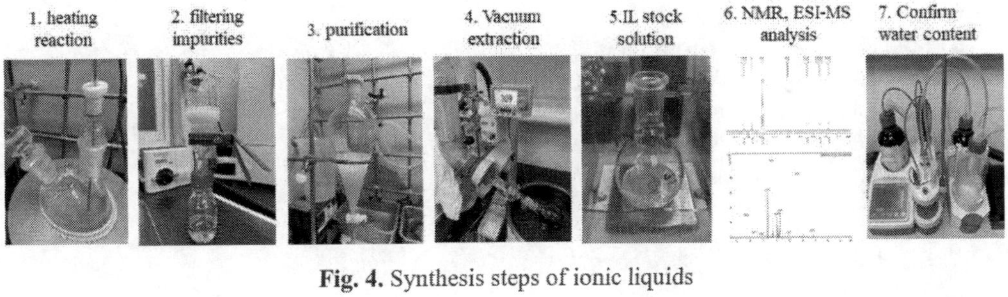

Fig. 4. Synthesis steps of ionic liquids

 https://doi.org/10.1051/e3sconf/202562002002

2.3 Dynamic absorption test of NH_3

The NH_3 absorption performance of ionic liquids was performed by a dynamic absorption testing system. Experiment diagram is shown in Fig. 5. In the gas absorption system, NH_3 (100,000 ppm balanced with N_2) and compressed air were used as an absorbate and carrier gas, respectively. The gas flow rate was controlled using a mass flow controller and the total flow rate of test gas was 6000 mL/min at room temperature and RH=45~50%. The ionic liquid was placed into an absorption reactor in a thermostatic bath to maintain the absorb temperature at 25 °C. Before the absorption test, NH_3 concentration was detected by an ammonia monitor (POLI, range: 0~500 ppm, resolutions: 1 ppm) through the bypass route. When the NH_3 concentration was stable, the gas was switched into the absorption reactor and started the absorption test.

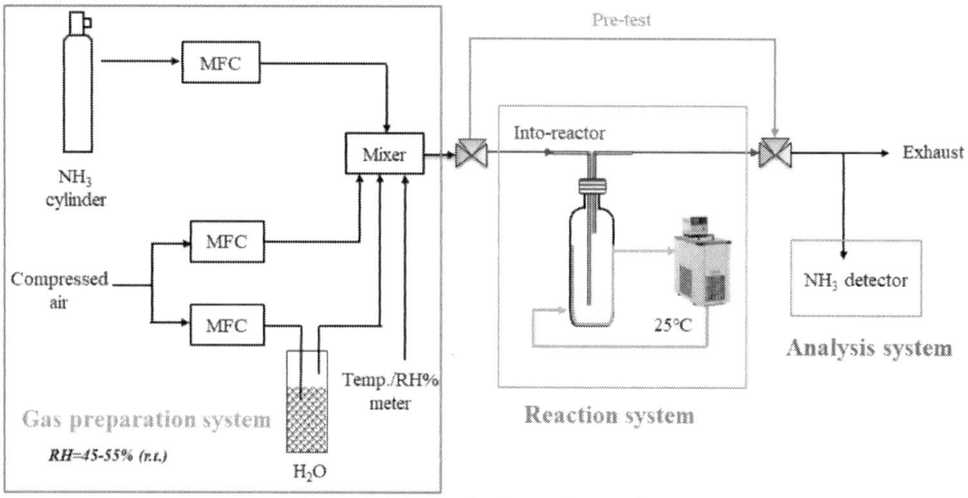

Fig. 5. NH_3 dynamic absorption testing system

The calculation method for NH_3 absorption capacity can be derived from the adsorption curve. The absorption capacity calculation method is as follows:

$$Q = C_{in} \times V \times \frac{1}{0.082} \times \frac{1}{(273 + T)} \times M \int \eta(t)dt \times 10^{-3}$$

where
Q: the absorption capacity (g),
C_{in}: the upstream gas concentration (ppm),
V: the test airflow volume (m^3/min),
T: the system temperature (°C),
M: the molecular weight of the standard test gas (g/mole),
η: the gas removal efficiency,
t: the test time.

 https://doi.org/10.1051/e3sconf/202562002002

3 Results and discussion

3.1 NH_3 Removal Efficiency (preliminary test)

The NH_3 removal efficiency of ionic liquids was tested by a low-concentration NH_3 dynamic absorption system. NH_3 was purged into the ionic liquids using an aeration method, and the compared standard was used with 90% removal efficiency. Six ionic liquids were synthesized in this study (final products shown in Fig. 6.). The synthesized ionic liquids were analyzed using NMR and ESI-MS, confirming that their structures matched the expected design. After verifying the structures of the ionic liquids, different concentrations were prepared for NH_3 removal efficiency testing.

Deionized water (D.I. water) and tap water were used as benchmarks, and the ionic liquid material IL-1 was used for preliminary testing. The test conditions were set with an NH_3 inlet concentration of 10 ppm, relative humidity of 50%, a total mixed gas flow rate of 3000 sccm, and an absorption liquid volume of 50 mL.

From the experimental results shown in Fig. 7 and Table 1, it is evident that using D.I. water and tap water as the absorption liquid results in poor NH_3 removal efficiency with NH_3 concentration breaking through quickly. The breakthrough times for D.I. water and tap water are 136 minutes and 102.5 minutes, respectively. In contrast, 80% IL-1 can maintain an NH_3 outlet concentration of 0 ppm for more than 1780 minutes, demonstrating higher absorption performance. However, the experiment duration was excessively long due to the high absorption efficiency of IL-1. To determine the breakthrough time and calculate the 90% removal efficiency more efficiently, the testing conditions were adjusted to accelerate experiments.

Fig. 1. Photograph of the ionic liquids

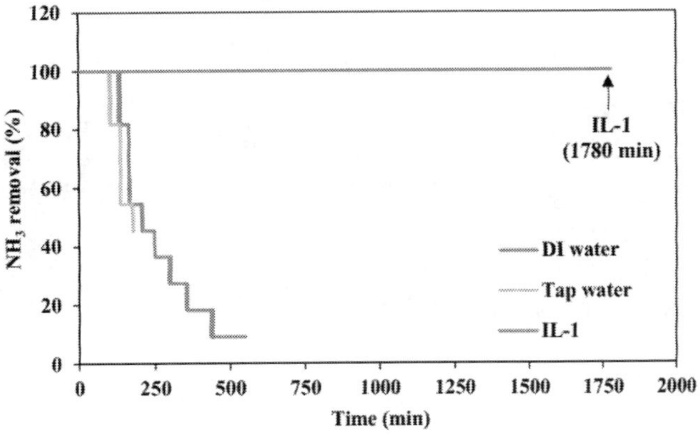

Fig. 7. NH_3 removal efficiency
(Total gas flowrate: 3000 sccm; NH_3 inlet conc.: 10 ppm; absorbent volume: 25 cm^3)

Table 1. Breakthrough time of different absorbent

	Material	Breakthrough time (mins)
Inlet NH_3 conc. = 10 ppm	DI water	136
	Tap water	102.5
	IL-1	>1780

* Test conditions: Inlet NH_3 conc.=10 ppm; Total gas flow=3000 sccm; RH=50±5% (r.t.); absorbent volume=25 cm^3.

3.2 NH_3 Removal Efficiency (first accelerate test)

The NH_3 inlet concentration was adjusted from 10 ppm to 100 ppm while keeping other conditions constant (relative humidity at 50% and mixed gas flow rate at 3000 sccm) to accelerate the experiment. The results are shown in Fig. 8 and Table 2. D.I. water was used as a benchmark again. Due to the tenfold increase in NH_3 inlet concentration, the NH_3 concentration broke through quickly, showing almost no removal efficiency. In contrast, 80% IL-1 could effectively absorb NH_3 for an extended period. NH_3 concentration breaks through only after 886 minutes which demonstrates excellent removal efficiency. However, the absorption testing time for IL-1 remained excessively long. To enable more efficient screening of ionic liquids for their absorption efficiency, a second phase of accelerated experiments was conducted, adjusting the test parameters to identify the optimal material.

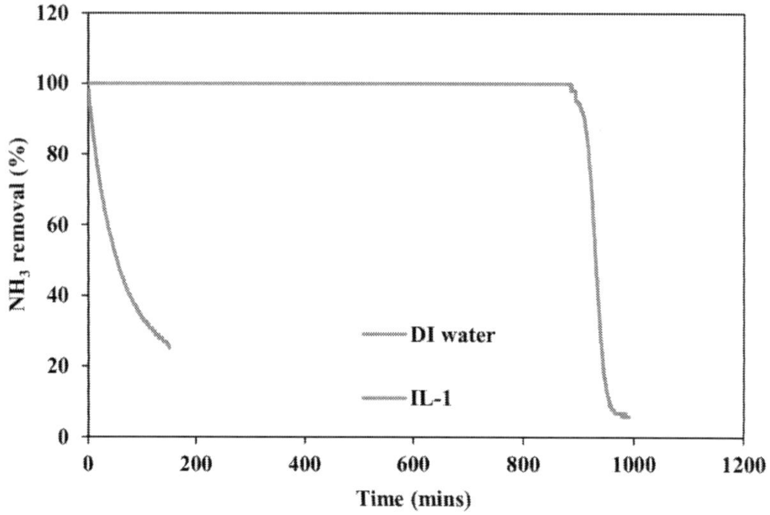

Fig. 2. NH_3 removal efficiency (Total gas flowrate: 3000 sccm; NH_3 inlet conc.: 100 ppm; absorbent volume: 25 cm^3)

Table 2. Breakthrough time of different absorbent (first accelerate test)

	Absorbent	**Breakthrough time (mins)**
Inlet NH_3 conc. = 100 ppm	DI water	0
	IL-1	886

*Test conditions: Inlet NH_3 conc.=100 ppm; Total gas flow=3000 sccm; RH=50±5% (r.t.); Absorbent volume=25 cm^3.

3.3 NH_3 Removal Efficiency (second accelerate test)

To solve the problem of low-concentration NH_3 emissions problem in semiconductor facilities (NH_3 concentrations in exhaust pipelines ranging from 5 to 100 ppm), the NH_3 inlet concentration was maintained at 100 ppm for the second accelerated experiment. The testing conditions were further adjusted by increasing the total mixed gas flow rate from 3000 sccm to 6000 sccm and reducing the absorption liquid volume from 50 mL to 25 mL. By increasing the liquid-to-gas ratio of the absorption reaction, the NH_3 breakthrough time was reduced.

The NH_3 removal efficiency trends for the various ionic liquids are shown in Figure 9, with the results detailed in Table 1. In terms of 90% removal efficiency absorption capacity, IL-1 and IL-4/IL-5 exhibit superior NH_3 absorption performance. Their absorption efficiency increases with the concentration of the ionic liquid, demonstrating a positive correlation between the quantity of ionic liquid in the absorption solution and NH_3 absorption efficiency. Additionally, comparing the absorption efficiency of different ionic liquids at 20% concentration, IL-5 shows the best performance, maintaining an outlet NH_3 concentration of 0 ppm for over 351 minutes, proving its excellent NH_3 absorption efficiency.

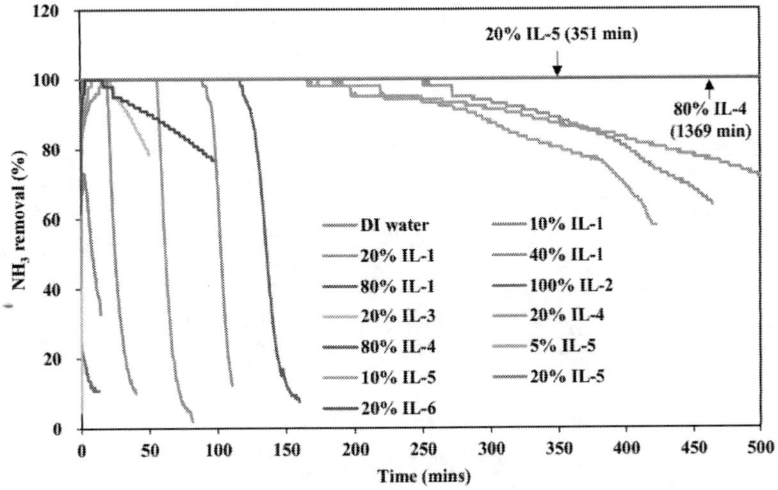

Fig. 3. NH_3 removal efficiency
(Total gas flowrate: 6000 sccm; NH_3 inlet conc.: 100 ppm; absorbent volume: 25 cm^3)

Table 3. NH_3 removal efficiency of ionic liquids*

Absorbent	Concentration	90% breakthrough time (min)	Solution weight (g)	I.L. weight (g)	90% absorption capacity (mg)
DI water	-	X	25	-	0
IL-1	10%	21	24.492	25	8.8
	20%	57.5	25.483	5.1	24.0
	40%	96.5	26.284	10.5	42.7
	80%	125	29.974	24.0	55.8
IL-2	100%	X	31.169	31.2	0
IL-3	20%	36	25.645	5.1	14.8
IL-4	20%	334.5	26.439	5.3	136.2
	80%	>1369	32.698	26.2	-
IL-5	5%	289.5	25.187	1.3	120.2
	10%	323	25.921	2.6	134.1
	20%	>351	27.186	5.4	-
IL-6	20%	49	25.785	5.2	19.5

*Test conditions：Inlet NH_3 conc.=100 ppm; Total gas flow=6000 sccm; RH=50±5% (r.t.); absorbent volume=25 cm^3.

References

1. TSMC Sustainability Report (2022)
2. Shaojuan, Zeng., Current Opinion in Green and Sustainable Chemistry (25), 100354 (2020)
3. Lei, Yuan., Journal of Molecular Liquids (323), 114601 (2021).
4. Stephen J. Osborne, The Royal Society of Chemistry (44), 11286 (2015).

Comparative Analysis of Alternative Sanitizers for Decontaminating Bacterial Counts on Aromatic Young Coconuts

Latchaporn Thamteerasathian[1*], *Kittayakarn Isarangura Na Ayuthya* [2], *Nattapong Veerataveeporn* [3], and *Pannarasi Susirirut* [4]

[1,3]Faculty of Business Administration, Chitralada Technology Institute. Bangkok, Thailand
[2]Office of General Education, Chitralada Technology Institute. Bangkok, Thailand
[4]Samroiyod Subdistrict Administration Organization, Prachuap Khiri Khan, Thailand

Abstract. The study explores alternative sanitizers to sodium hypochlorite (NaOCl) for reducing bacteria on aromatic young coconuts. Acetic acid, hydrogen peroxide (H_2O_2), and trisodium phosphate (Na_3PO_4) were evaluated for their antimicrobial efficacy. The results showed that acetic acid at 5% concentration for 30 minutes achieved the highest log reduction of 3.498, followed by H_2O_2 at 5% concentration for 5 minutes with a reduction of 2.814. Trisodium phosphate showed modest efficacy. The study also found that combined treatments, such as H_2O_2 followed by acetic acid, resulted in significant log reductions. All treatments showed statistically significant differences compared to the control. The findings suggest that acetic acid and hydrogen peroxide are effective alternatives to NaOCl for postharvest microbial control on aromatic young coconuts. The study highlights the importance of optimizing sanitizer concentration, exposure time, and combined treatments to maximize antimicrobial efficacy while maintaining product quality. Overall, the research provides insights into alternative sanitizers for the food industry, particularly for postharvest treatment of aromatic young coconuts.

1 Introduction

The global coconut water market is experiencing significant growth driven by rising health consciousness, increased demand for plant-based diets, and the popularity of natural and hydrating beverages. The coconut water market was valued at approximately USD 7.5 billion in 2023 and is expected to grow at a compound annual growth rate (CAGR) of 14.4%, reaching USD 26.1 billion by 2032 [1, 2].

Coconuts are significant globally and within Thailand [3, 4]. In 2023, Thailand's exports of coconut water reached significant values. Specifically, the value of coconut water exports from Thailand to various countries amounted to around $30 million in November 2023. A substantial portion of these exports, approximately $25 million, was directed to China, the largest market for Thailand's coconut water, accounting for about 85% of the total export

* Corresponding author: latchaporn.tha@cdti.ac.th

value. The United States was the second-largest market, with exports valued at around $2 million [5].

Aromatic young coconut (fragrant coconut or Maphrao Num Hom) is an agricultural commodity that Thailand has the potential to produce, with a growing export trend. This variety originated from a mutation of Moo See, a dwarf aromatic water coconut. The fruit, grown from *Cocos nucifera* L. of the Arecaceae family, is cultivated in tropical and subtropical regions worldwide [6, 7].

The modern consumer's desire for nutritious, healthy, natural foods with high organoleptic leads to the demand for minimally processed foods, including coconut water [8, 9]. Fresh fruits and vegetables, including coconut, are significant components of the global food market, providing dietary fibre, vitamins, minerals, phytochemicals, antioxidants, and organic acids with health benefits [10, 11].

Coconut water, often consumed from immature fruits, is a refreshing drink valued for its sweetness and natural appeal. It is gaining attention for its natural functional properties and becoming a popular healthy beverage with potential therapeutic properties, such as preventing and relieving dehydration, constipation, digestive problems, fatigue, heatstroke, diarrhoea, kidney stones, and urinary infections [3, 4, 12]. Coconut's unique characteristics, including its sugar content and mineral composition, make it an ideal natural alternative to commercial sports drinks, rehydrating and refreshing drink post-exercise due to its electrolyte-replenishing properties, including low acidity and isotonic mineral contents [3, 4, 12]. It is favoured for its pleasurable scent and sweet taste, which come from various parts of the coconut, such as water and meat [7]. Fresh coconut water is also rich in cytokinins, which can delay the onset of aging characteristics in human skin cells, potentially providing anti-aging effects [12].

The shelf life of immature coconuts can be prolonged with postharvest treatments. For commercializing coconut water as a ready-to-drink beverage, adequate heat treatment is essential to prevent microbial and oxidative enzymatic spoilage and also to preserve the original flavour of coconut water, which is sensitive to temperature [3, 12]. Minimally processed methods can help maintain its aroma. Minimally processed foods, subjected to washing, sanitizing, peeling, and cutting, are increasingly popular but more perishable and vulnerable to spoilage than whole produce [13, 14].

Fresh produce, including minimally processed fruits and vegetables, often has a high initial microbial load, necessitating washing and sanitizing to enhance safety and prolong shelf life [10, 14]. Simple washing with water is insufficient, and chemical sanitizers like chlorine dioxide, hydrogen peroxide, organic acids, and ozone are used to reduce microbial load to acceptable levels [8, 15].

Although widely used for food sanitization, chlorination cannot eradicate all germs on fresh vegetables. Chemical reactions between chlorine and organic matter produce toxic by-products and residues on produce, reducing its disinfection efficacy [10].

Recent outbreaks raise doubt about chlorine's efficacy and environmental hazards, including the development of carcinogenic disinfection by-products (DBPs) such as haloacetic acids (HAAs) and trihalomethanes (THMs). These chlorine compounds and sodium hypochlorite by-products may irritate the skin and lungs [16, 17]. These issues motivated organic acids and other non-chlorine disinfection studies [18, 19].

Other studied alternative sanitizers include trisodium phosphate, organic acids, hydrogen peroxide, and ozone [15]. Their efficacy depends on microbial load and environmental conditions [20]. The factors for choosing alternative chemical sanitizers are safe, cost-effective, water-applicable, flexible, residue-free, and non-selectively toxic, increasing their commercial demand [10].

The objective of this study was to compare the efficacy of various alternative sanitizers to sodium hypochlorite in reducing the initial total aerobic mesophilic bacteria on the surfaces of aromatic young coconuts (*Cocos nucifera* L.).

2 Method and Materials

2.1 Aromatic coconuts preparation

Thai aromatic young coconut (*Cocos nucifera* L.) cv. Nam Hom, at 20-25 days old, estimated a soft, thin layer of coconut flesh. The average weight was 1.6-2.1 kg. The coconuts were purchased from the same areas and the same local producers in Ban Phaeo, Samut Sakhon province, Thailand. The coconuts were experimented with no later than 12 hours after harvesting.

Fifty aromatic young coconuts per the main experiment, free of visual defects, were used in the study. The surface of the coconut skin was washed by immersing it in tap water, draining it, and air-drying it. The coconut was trimmed and cut, and part of the green exocarp was cut (only the top was trimmed into the oval shape (this part is for perforation to get coconut water for further processing).

Aromatic young coconuts were categorized into six main sanitizer groups with sub-groups for each sanitizer, as shown in Fig. 1 and Table 1.

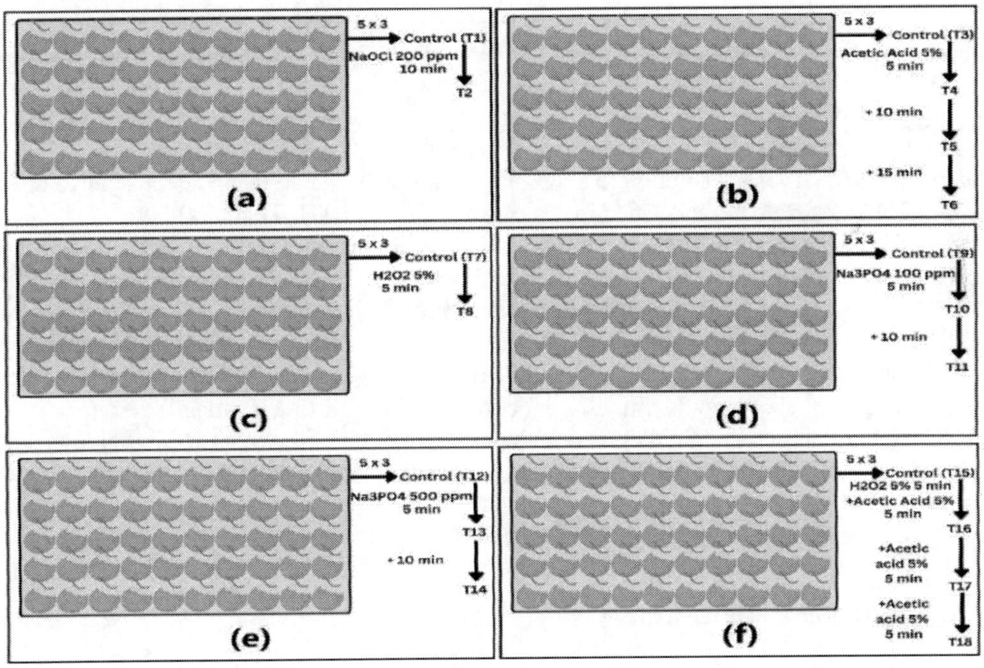

Fig. 1. Illustration of decontamination treatments (a) Treatment of NaOCl 200 ppm

(b) Treatment of acetic acid 5% (w/w) (c) Treatment of H2O2 5% (w/w) (d) Treatment of Na3PO4 100 ppm (e) Treatment of Na3PO4 500 ppm (f) Treatment of H2O2 5% (w/w) 5 minutes followed by acetic acid 5% (w/w)

Table 1. Decontamination treatments

Sanitizing Agent	Sampling coconuts (n) x Replication	Addition process from prior treatment	Treatment No.
Control*-(a)	(5) x 3	-	T1
NaOCl 200 ppm		10 minutes	T2
Control-(b)	(5) x 3	-	T3
Acetic acid 5% (w/w)		5 minutes	T4
		+ 5 minutes	T5
		+ 5 minutes	T6
Control (c)	(5) x 3	-	T7
H_2O_2 5% (w/w)		5 minutes	T8
Control-(d)	(5) x 3	-	T9
Na_3PO_4 100 ppm		5 minutes	T10
		+10 minutes	T11
Control-(e)	(5) x 3	-	T12
Na_3PO_4 500 ppm		5 minutes	T13
		+10 minutes	T14
Control-(f)	(5) x 3	-	T15
H_2O_2 5 % (w/w) 5 minutes followed by acetic acid 5% (w/w)		Acetic acid 5% (w/w) 5 minutes	T16
		Acetic acid 5% (w/w) +5 minutes	T17
		Acetic acid 5% (w/w)+5 minutes	T18

2.2 Materials and procedures

Acetic acid (CH3COOH) solution (5% w/w), which was prepared using acetic acid solution (Glacial, 99.85%, Food grade, Chemipan Corporation Co., LTD, Thailand)

Sodium hypochlorite (NaOCl) solution of 200 ppm (parts per million) free chlorine was prepared using Hi-Chlor (active chlorine 10%, Food grade, ACG Chemicals (Thailand) Co., Ltd, Thailand). The pH was adjusted to 6.5 using a citric acid solution. The free chlorine level in the sodium hypochlorite solution was determined by titration with sodium thiosulfate.

Hydrogen peroxide (H2O2) solution (5%), which was prepared using 6% hydrogen peroxide solution (Food grade, Enrich Fogger (Thailand) Co., Ltd, Thailand).

Trisodium phosphate (Na3PO4) solution (100 and 500 ppm), which was prepared using trisodium phosphate granular (Food grade, Aditya Birla Chemicals (Thailand) Ltd, Thailand).

Sanitizing solutions were prepared immediately before use.

2.3 Decontamination treatment

The swab test for 5x5 cm^2 was used to determine the aerobic mesophilic bacteria, which were expressed as CFU/cm^2.

Initial contamination levels were determined with five aromatic young coconuts and three replications swabbed on the surfaces before sanitizing treatments were used as a control.

Aromatic young coconuts were dipped in specified sanitizers at ambient temperature according to Table 1 and rinsed with sterile water. Three replicates were carried out for each main experiment, which was performed five times.

Samples were immersed in the following sanitizer solutions: (a) Control as T1, the coconuts were continued immersing in NaOCl 200 ppm for 10 minutes as T2; (b) Control as

T3, the coconuts were continued immersing in acetic acid 5% (w/w) for 5 minutes as T4, the coconuts were continued immersing in acetic acid 5% (w/w) for more 10 min (total time was 15 minutes) as T5, and the coconuts were continued immersing in acetic acid 5% (w/w) for more 15 minutes (total time was 30 minutes) as T6; (c) Control as T7, the coconuts were continued immersing in H_2O_2 5% (w/w) for 5 minutes as T8; (d) Control as T9, the coconuts were continued immersing in Na_3PO_4 100 ppm for 5 minutes as T10, the coconuts were continued immersing in Na_3PO_4 100 ppm for more 10 minutes (total time was 15 minutes) as T11; (e) Control as T12, the coconuts were continued immersing in Na_3PO_4 500 ppm for 5 minutes as T13, the coconuts were continued immersing in Na_3PO_4 500 ppm for more 10 minutes (total time was 15 min) as T14; (f) Control as T15, the coconuts were continued immersing in H_2O_2 5% (w/w) for 5 minutes and followed by acetic acid 5% (w/w) for more 5 minutes as T16, the coconuts were continued immersing in acetic acid 5% (w/w) for more 5 minutes (total time was 10 min) as T17, the coconuts were continued immersing in acetic acid 5% (w/w) for more 5 minutes (total time was 15 minutes) as T18. All treatment was rinsed with sterile water after decontamination treatment.

These washing treatments were tested for their ability to reduce bacterial populations on the aromatic young coconut surface.

2.4 Microbiological analysis

The samples were subjected to bacteriological analyses to determine initial contamination levels, surviving microflora counts after washing in tap water (as a control), and microbial counts after various sanitizing agents by swab test.

Each treatment was swabbed on different surface areas for 5 x 5 cm^2 with a sterile disposable cotton swab, where the swab was transferred to a tube containing 10 mL of peptone water. Then, the sample was serially diluted with phosphate buffer (pH 7.2).

The plate count agar medium used in this study was purchased from M & P Impex Limited Partnership, Thailand (BD Difco™ & BBL™, Sparks, MD, USA).

Total aerobic mesophilic bacteria counts were determined by pouring 0.1 mL of diluted sample into plate count agar and using the spread plate technique. Plates were incubated at 35 °C for 48 hours, and colonies were counted and expressed as log CFU/cm^2 of aromatic young coconut surface.

All plates used for quantitative analysis were carried out in three replicates at each relevant dilution.

Sanitizing solutions were prepared immediately before use.

2.5 Statistical analysis

Microbial data determined from aromatic young coconut surfaces were converted to log CFU/cm^2. Each treatment was repeated three times, with five coconuts per trial. All data was calculated and expressed as mean ± standard deviation of triplicate measurements and analysed by applying the one-way analysis of variance (ANOVA). Duncan's multiple range test was used for pairwise comparison. Significance was defined as $p \leq 0.05$ between mean total aerobic bacterial count levels for different treatments.

These values were analysed for differences in response to the efficacy of each treatment.

3 Results and Discussion

3.1 Effect of various sanitizing agents for decontamination initial aerobic mesophilic bacteria on aromatic young coconut (*Cocos nucifera* L.) surfaces.

(a) Sodium hypochlorite (NaOCl)

The study on the reduction of aerobic mesophilic bacteria on the surface of aromatic young coconuts using sodium hypochlorite (NaOCl) at a concentration of 200 ppm for 10 min demonstrated a significant decrease in bacterial load. The initial bacterial count was 4.207 ± 0.484 log CFU/cm^2, which was reduced to 1.714 ± 0.516 log CFU/cm^2, resulting in a log reduction of 2.493 log CFU/cm^2 as shown in Table 2. The log reduction observed in this study aligns with previous research findings demonstrating the antimicrobial efficacy of NaOCl at similar concentrations. A study by Petri et al. (2021) reported that treatment with NaOCl at 150 ppm for 15 minutes reduced the initial contamination by 2-3 log cycles in minimally processed lettuce. This finding indicates that even slightly lower concentrations of NaOCl can achieve comparable reductions, though exposure time plays a critical role in determining overall effectiveness [16].

Conversely, Beuchat et al. (1998) highlighted the variability in NaOCl efficacy, reporting that 200 ppm NaOCl for 10 minutes was less effective in reducing aerobic loads on apples, tomatoes, and lettuce, achieving reductions of 1.94, 1.74, and 0.33 log CFU/cm2, respectively [15]. These discrepancies underscore the influence of produce type, surface characteristics, and organic matter presence on the antimicrobial action of NaOCl.

Moreover, the synergistic effects of combining NaOCl with other treatments have been documented to enhance its bactericidal properties. For instance, 100 reported that using ultrasound with NaOCl (100 mg/L) for 5 minutes on arugula leaves enhanced bactericidal effects, reducing microbial contamination by 1.46 log cycles compared to 0.85 log cycles with NaOCl alone. This suggests that innovative combinations of sanitizing methods can improve the effectiveness of NaOCl without significantly altering the physicochemical properties of the treated produce.

Given the adverse effects of using NaOCl, such as the potential formation of harmful by-products, there is growing interest in alternative sanitizing agents for reducing microbial contamination in fruits and vegetables.

(b) Acetic acid

The study on the antimicrobial efficacy of acetic acid on the surface of aromatic young coconuts reveals significant reductions in aerobic mesophilic bacteria counts with varying exposure times. Initially, the bacterial load was 4.673 ± 0.314 log CFU/cm^2, which decreased to 2.577 ± 0.311 log CFU/cm^2 after 5 minutes of immersion in 5% acetic acid, resulting in a log reduction of 2.096 log CFU/cm^2. Extending the immersion time to 15 min further reduced the bacterial count to 2.371 ± 0.306 log CFU/cm^2, achieving a log reduction of 2.303 log CFU/cm^2. The most substantial reduction was observed with 30 minutes of immersion, where the bacterial count dropped to 1.175 ± 0.322 log CFU/cm^2, corresponding to a log reduction of 3.498 log CFU/cm^2, as shown in Table 2. The highest log reduction observed in this study was 5% acetic acid for 30 minutes, with 3.498 log reductions in aerobic mesophilic bacteria.

This reduction was higher than the treatment of sodium hypochlorite 100 ppm for 10 minutes. As a result, treatment with 5% acetic acid for 30 minutes is an excellent alternative sanitizing agent to decontaminate aerobic mesophilic bacteria. Still, slight drawbacks of pungent odour and distinct taste can affect the sensitive produces' sensory qualities. The observed increase in log reduction with extended exposure time highlights the importance of treatment duration in maximizing the antimicrobial effects of acetic acid. The substantial reduction of 3.498 log CFU/cm^2 achieved after 30 minutes of immersion indicates that prolonged exposure significantly enhances the bactericidal properties of acetic acid.

In comparison, Nascimento et al. (2003) found that dipping lettuce samples in 2% acetic acid for 15 minutes resulted in a log reduction of 3.37 log CFU/g, while 4% acetic acid achieved a reduction of 3.91 log CFU/g for aerobic mesophilic populations [21]. Our study's log reduction of 2.303 log CFU/cm^2 after 15 minutes with 5% acetic acid is slightly lower than these findings. This discrepancy could be due to differences in produce surface morphology, initial microbial load, and concentration of acetic acid used.

(c) Hydrogen peroxide (H_2O_2)

The study on the reduction of aerobic mesophilic bacteria on the surface of aromatic young coconuts using 5% hydrogen peroxide (H_2O_2) for 5 minutes demonstrated a significant decrease in bacterial count from an initial level of 4.979 ± 0.285 log CFU/cm^2 to 2.166 ± 0.216 log CFU/cm^2, resulting in a log reduction of 2.814 log CFU/cm^2 as shown in Table 2. A study by Alexandre et al. (2012) observed that using 5% H_2O_2 for 2 minutes on strawberries resulted in a 2.25 log reduction of total mesophiles [22]. This trend is consistent with our findings, although our exposure time was longer (5 minutes), achieving a higher log reduction of 2.814 log CFU/cm 2. This suggests that longer exposure times enhance the effectiveness of H_2O_2.

Additionally, Sapers et al. (2001) found that the log reduction was higher when 5% H_2O_2 was applied at elevated temperatures, achieving log reductions of 3 or higher when commodities such as apples and cantaloupes were fully immersed in H_2O_2 solutions with vigorous agitation at temperatures of 50-60 °C and 70-80 °C, respectively; further supporting our findings, they reported that population reductions on washed rind plugs were less than 1 log with water, 1 to 2 logs with washing and sanitizing agents applied individually, and up to 3 logs with some sequential treatments involving H_2O_2, with H_2O_2 applied at 50 °C being superior to other treatments, yielding a fresh-cut shelf-life of over 2 weeks, which underscores the potential for optimized application conditions to enhance the effectiveness of H_2O_2, as evidenced by our observed log reduction of 2.814 log CFU/cm^2 [23].

Lastly, CDC (2008) highlighted the bactericidal properties of a 7% H_2O_2 solution, achieving more than 6 log reductions in 5 minutes when tested after 14 days of stress. Although our study utilized a lower concentration (5%) and achieved a lower log reduction (2.814 log CFU/cm^2), these findings indicate the strong antimicrobial potential of H_2O_2 and suggest that higher concentrations could further improve efficacy [24].

While stabilized H_2O_2 was 3-4 times more effective than sodium hypochlorite, plain H2O2 was slightly less effective than stabilized hydrogen peroxide [25].

(d) and (e) Trisodium phosphate (Na_3PO_4)

The study on the effectiveness of trisodium phosphate (Na_3PO_4) in reducing aerobic mesophilic bacteria on the surface of aromatic young coconuts provides insightful data on microbial control in postharvest treatments. Initially, the average levels of viable mesophilic bacteria on the coconut surface were 4.132 ± 0.122 log CFU/cm^2. After immersing the coconuts in 100 ppm Na_3PO_4 for 5 and 15 minutes, the bacterial counts reduced to 4.004 ± 0.126 and 3.851 ± 0.125 log CFU/cm^2, respectively, corresponding to log reductions of 0.128 and 0.281. When the concentration of Na_3PO_4 was increased to 500 ppm, the initial bacterial count of 4.758 ± 0.148 log CFU/cm^2 decreased to 4.251 ± 0.149 and 4.123 ± 0.145 log CFU/cm^2 after 5 and 15 minutes, respectively, with log reductions of 0.506 and 0.635 as shown in Table 2. The results of this study show modest log reductions in viable mesophilic bacteria with the use of Na_3PO_4, with greater effectiveness observed at higher concentrations.

The log reductions achieved with 500 ppm Na_3PO_4 (0.506 after 5 minutes and 0.635 after 15 minutes) indicate a dose-response relationship, where increasing the concentration of the sanitizing agent enhances its antimicrobial efficacy.

In a related study, Sapers et al. (2001) observed a 1.33 log reduction in total bacterial count on cantaloupe melons washed for 10 minutes with 4% Na_3PO_4 and a 1.13 log reduction with 8% Na_3PO_4 [23]. They also reported a log reduction of 1.81 log CFU/cm^2 when 5%

Na_3PO_4 was combined with 5% H_2O_2 and rinsed, compared to individual reductions of 1.33 log CFU/cm^2 with 4% Na_3PO_4 and 1.65 log CFU/cm^2 with 5% H_2O_2 alone, suggesting that combined treatments can synergistically enhance antimicrobial efficacy. Similarly, Liao & Sapers (2000) reported a 1 log reduction on the surface of tomatoes when dipped in 15% Na_3PO_4 for 15 seconds, indicating that even short exposure times can be effective at high concentrations [26]. These findings support the observation that the antimicrobial efficacy of sodium phosphate increases with concentration.

(f) Combination of H_2O_2 and acetic acid

The study's findings on reducing aerobic mesophilic bacteria on the surface of aromatic young coconuts through immersion in hydrogen peroxide (H_2O_2) and acetic acid reveal significant insights into effective sterilization techniques. Initially, the bacterial load was 4.687 ± 0.116 log CFU/cm^2, which was reduced to 2.057 ± 0.124 log CFU/cm^2 after 5 minutes immersion in 5% H_2O_2 followed by 5 minutes immersion in 5% acetic acid, achieving a log reduction of 2.631. Extending the immersion in acetic acid to 10 and 15 minutes further reduced the bacterial counts to 1.832 ± 0.125 and 1.753 ± 0.129 log CFU/cm^2, respectively, with log reductions of 2.856 and 2.934. These results underscore the efficacy of combined H_2O_2 and acetic acid treatments in significantly lowering bacterial contamination on coconut surfaces

In conclusion, this study identifies alternative sanitizing agents to sodium hypochlorite (NaOCl), which achieved a log reduction of 2.493 log CFU/cm^2 at 200 ppm for 10 minutes. Acetic acid (5% w/w) for 30 minutes resulted in the highest log reduction of 3.498 log CFU/cm^2, making it the most effective alternative. This was followed by hydrogen peroxide (H_2O_2) at 5% for 5 minutes and acetic acid (5% w/w) for 15 minutes, with log reductions of 2.934 log CFU/cm^2 and 2.856 log CFU/cm^2, respectively. These treatments showed statistically significant differences ($p \leq 0.05$), as shown in Fig. 3.

The treatment of H_2O_2 at 5% for 5 minutes achieved a log reduction of 2.814 log CFU/cm^2, which was not significantly different from the combined treatments of H_2O_2 at 5% for 5 minutes and acetic acid at 5% for 15 minutes, and H_2O_2 at 5% for 5 minutes and acetic acid at 5% for 10 minutes. The least effective treatments compared to NaOCl were H_2O_2 at 5% for 5 minutes, followed by acetic acid at 5% for 5 minutes, with a log reduction of 2.631 log CFU/cm^2 ($p \leq 0.05$), as shown in Fig. 3.

The treatments were less effective in log reduction than sodium hypochlorite at 200 ppm for 10 minutes, which reduced 2.493 log CFU/cm^2. Specifically, 5% (w/w) acetic acid for 15 minutes achieved a reduction of 2.303 log CFU/cm^2, 5% (w/w) acetic acid for 5 minutes achieved 2.096 log CFU/cm^2, 500 ppm Na_3PO_4 for 15 minutes achieved 0.635 log CFU/cm^2, 500 ppm Na_3PO_4 for 5 minutes achieved 0.506 log CFU/cm^2, 100 ppm Na_3PO_4 for 15 minutes achieved 0.281 log CFU/cm^2, and 100 ppm Na_3PO_4 for 5 minutes achieved 0.128 log CFU/cm^2. All treatments showed statistically significant differences ($p \leq 0.05$), as shown in Fig. 3.

Table 2. The log reductions after treatment with water and various sanitizing agents

Treatment	Mean Log* CFU**/cm^2	SD	Mean Log Reduction
Control-NaOCl	4.207	0.484	0.000
NaOCl 200 ppm 10 minutes	1.714	0.516	2.493^{e}
Control-Acetic Acid 5% (w/w)	4.673	0.314	0.000
Acetic Acid 5% (w/w) 5 minutes	2.577	0.311	2.096^{g}
Acetic Acid 5% (w/w) 15 minutes	2.371	0.306	2.303^{f}
Acetic Acid 5% (w/w) 30 minutes	1.715	0.322	3.498^{a}
Control-H_2O_2 5% (w/w)	4.979	0.285	0.000
H_2O_2 5% (w/w) 5 minutes	2.166	0.284	2.814^{bc}
Control-Na_3PO_4 100 ppm	4.132	0.122	0.000

Treatment	Mean Log* CFU**/cm^2	SD	Mean Log Reduction
Na_3PO_4 100 ppm 5 minutes	4.004	0.126	0.128^k
Na_3PO_4 100 ppm 15 minutes	3.851	0.125	0.281^j
Control-Na_3PO_4 500 ppm	4.758	0.148	0.000
Na_3PO_4 500 ppm 5 minutes	4.251	0.149	0.506^i
Na_3PO_4 500 ppm 15 minutes	4.123	0.145	0.635^h
Control-H2O2 5% (w/w) and Acetic Acid 5% (w/w)	4.687	0.116	0.000
H2O2 5% (w/w) 5 min followed by Acetic acid 5% (w/w) 5 minutes	2.057	0.124	2.631^d
H2O2 5% (w/w) 5 min followed by Acetic acid 5% (w/w) 10 minutes	1.832	0.125	2.856^c
H2O2 5% (w/w) 5 min followed by Acetic acid 5% (w/w) 15 minutes	1.753	0.129	2.934^b

Remark:
*All counts are expressed in logarithms of reduction in aerobic mesophilic bacteria counts.
*Mean and standard deviations (SD) represent five samples per experiment and three experiments for each treatment. Values with different letters in the same column are significantly different ($p \leq 0.05$; Duncan's multiple range test).
**CFU: colony-forming unit.

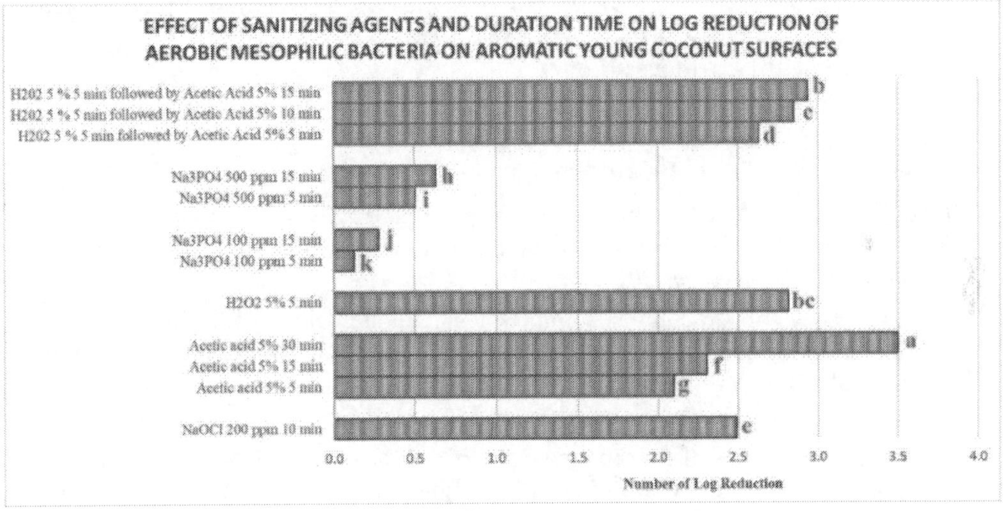

Fig. 3. Effect of sanitizing agents and duration time on log reduction of aerobic mesophilic bacteria on aromatic young coconut surfaces

Remark: Values with different letters at the end of the column are significantly different

($p \leq 0.05$; Duncan's multiple range test)

4 Conclusion

The study demonstrates the effectiveness of various sanitizing agents as alternatives to sodium hypochlorite in reducing aerobic mesophilic bacteria on the surfaces of aromatic young coconuts, highlighting the superior efficacy of acetic acid and hydrogen peroxide under specific conditions and significant conclusions are as follows:

1. Acetic acid at 5% (w/w) for 30 minutes emerged as the most effective alternative to NaOCl, achieving the highest log reduction in bacterial load.
2. Hydrogen peroxide at 5% (w/w) for 5 minutes and combined treatments with acetic acid were also highly effective, significantly reducing bacterial contamination.
3. Trisodium phosphate showed moderate efficacy, with more significant reductions at higher.

These results indicate that while sodium hypochlorite remains effective, acetic acid and hydrogen peroxide offer promising microbial control alternatives in postharvest aromatic young coconut treatments.

This study was supported by Chitralada Technology Institute. Bangkok, Thailand.

References

1. IMARC Services Private Limited. (2024). *Coconut Water Market Size, Share, Industry Analysis 2022-27*. Www.imarcgroup.com. https://www.imarcgroup.com/coconut-water-market
2. Mordor Intelligence. (2024). *coconut water Market | 2022 - 27 | Industry Share, Size, Growth - Mordor Intelligence*. Www.mordorintelligence.com. https://www.mordorintelligence.com/industry-reports/coconut-water-market
3. Prades, A., Dornier, M., Diop, N., & Pain, J. P. (2012a). Coconut water preservation and processing: A review. *Fruits*, *67*(3), 157–171. https://doi.org/10.1051/fruits/2012009
4. Prades, A., Dornier, M., Diop, N., & Pain, J. P. (2012b). Coconut water uses, composition and properties: A review. *Fruits*, *67*(2), 87–107. https://doi.org/10.1051/fruits/2012002
5. IndexBox. (2024). *Thailand's Coconut Exports Reach $30M in November 2023 - News and Statistics - IndexBox*. Www.indexbox.io. https://www.indexbox.io/blog/thailand-coconut-exports-november-2023/
6. National Bureau of Agricultural Commodity and Food Standards Ministry of Agriculture and Cooperatives. (2007). Thai agricultural standard: Aromatic coconut (TAS 15-2007). In *https://www.acfs.go.th/standard/download/eng/coconuts.pdf*.
7. Mahayothee, B., Koomyart, I., Khuwijitjaru, P., Siriwongwilaichat, P., Nagle, M., & Müller, J. (2016). Phenolic compounds, antioxidant activity, and medium chain fatty acids profiles of coconut water and meat at different maturity stages. *International Journal of Food Properties*, *19*(9), 2041–2051. https://doi.org/10.1080/10942912.2015.1099042
8. Sethi, S., Nayak, S. L., Joshi, A., & Sharma, R. R. (2020). Sanitizers for fresh-cut fruits and vegetables. *Fresh-Cut Fruits and Vegetables*, 99–119. https://doi.org/10.1016/b978-0-12-816184-5.00005-7
9. Zoellner, C., Aguayo-Acosta, A., Siddiqui, M. W., & Dávila-Aviña, J. E. (2018). Peracetic acid in disinfection of fruits and vegetables. In M. W. Siddiqui (Ed.), *Postharvest Disinfection of Fruits and Vegetables* (pp. 53–66). Andre Gerhard Wolff. https://doi.org/10.1016/b978-0-12-812698-1.00002-9
10. Sahoo, S. K., Tomar, M. S., & Pradhan, R. C. (2021). Disinfecting agents for controlling fruits and vegetable diseases after harvest. In *Food Losses, Sustainable Postharvest and Food Technologies* (pp. 103–151). https://doi.org/10.1016/b978-0-12-821912-6.00007-9

11. Foda, A., Ammar, M., Fouad, O., & Abou-Zaid. (2022). Using of some minimal processing techniques in preservation of some fruit juices.
12. Tan, T. C., Cheng, L. H., Bhat, R., Rusul, G., & Easa, A. M. (2014). Composition, physicochemical properties and thermal inactivation kinetics of polyphenol oxidase and peroxidase from coconut (*Cocos nucifera*) water obtained from immature, mature and overly-mature coconut. *Food Chemistry*, *142*, 121–128. https://doi.org/10.1016/j.foodchem.2013.07.040
13. Waghmare, R. B., & Annapure, U. S. (2017). Effects of hydrogen peroxide, modified atmosphere and their combination on quality of minimally processed cluster beans. *Journal of Food Science and Technology*, *54*(11), 3658–3665. https://doi.org/10.1007/s13197-017-2827-x
14. Vandekinderen, I., Van Camp, J., De Meulenaer, B., Veramme, K., Denon, Q., Ragaert, P., & Devlieghere, F. (2007). The effect of the decontamination process on the microbial and nutritional quality of fresh-cut vegetables. *Acta Horticulturae*, *746*, 173–180. https://doi.org/10.17660/actahortic.2007.746.19
15. Beuchat, L. R., Nail, B. V., Adler, B. B., & Clavero, M. R. S. (1998). Efficacy of Spray Application of Chlorinated Water in Killing Pathogenic Bacteria on Raw Apples, Tomatoes, and Lettuce. *Journal of Food Protection*, *61*(10), 1305–1311. https://doi.org/10.4315/0362-028x-61.10.1305
16. Petri, E., Virto, R., Mottura, M., & Parra, J. (2021). Comparison of Peracetic Acid and Chlorine Effectiveness during Fresh-Cut Vegetable Processing at Industrial Scale. *Journal of Food Protection*, *84*(9), 1592–1602. https://doi.org/10.4315/jfp-20-448
17. Alvaro, J. E., Moreno, S., Dianez, F., Santos, M., Carrasco, G., & Urrestarazu, M. (2009). Effects of peracetic acid disinfectant on the postharvest of some fresh vegetables. *Journal of Food Engineering*, *95*(1), 11–15. https://doi.org/10.1016/j.jfoodeng.2009.05.003
18. Vandekinderen, I., Devlieghere, F., De Meulenaer, B., Ragaert, P., & Van Camp, J. (2009). Optimization and evaluation of a decontamination step with peroxyacetic acid for fresh-cut produce. *Food Microbiology*, *26*(8), 882–888. https://doi.org/10.1016/j.fm.2009.06.004
19. Walter, E. H. M., Nascimento, M. S., & Kuaye, A. Y. (2009). Efficacy of sodium hypochlorite and peracetic acid in sanitizing green coconuts. *Letters in Applied Microbiology*, *49*(3), 366–371. https://doi.org/10.1111/j.1472-765x.2009.02670.x
20. Stearns, R., Freshour, A., & Shen, C. (2022). Literature review for applying peroxyacetic acid and/or hydrogen peroxide to control foodborne pathogens on food products. *Journal of Agriculture and Food Research*, *10*, 100442. https://doi.org/10.1016/j.jafr.2022.100442
21. Nascimento, M. S., Silva, N., Catanozi, M. P. L. M., & Silva, K. C. (2003). Effects of Different Disinfection Treatments on the Natural Microbiota of Lettuce. *Journal of Food Protection*, *66*(9), 1697–1700. https://doi.org/10.4315/0362-028x-66.9.1697
22. Alexandre, E. M. C., Brandão, T. R. S., & Silva, C. L. M. (2012). Assessment of the impact of hydrogen peroxide solutions on microbial loads and quality factors of red bell peppers, strawberries and watercress. *Food Control*, *27*(2), 362–368. https://doi.org/10.1016/j.foodcont.2012.04.012
23. Sapers, G. M., Miller, R. L., Pilizota, V., & Mattrazzo, A. M. (2001). Antimicrobial Treatments for Minimally Processed Cantaloupe Melon. *Journal of Food Science*, *66*(2), 345–349. https://doi.org/10.1111/j.1365-2621.2001.tb11344.x

24. CDC. (2008). *Disinfection and sterilization guideline*. Infection Control. https://www.cdc.gov/infection-control/hcp/disinfection-and-sterilization/index.html
25. Hassan, F., Lakshmanan, P., Geethalakshmi, V., & Mukundan, M. (2013). Evaluation of stabilised hydrogen peroxide as sanitiser in seafood processing industry. *Indian Journal of Fisheries*, *60*(2).
26. Liao, C. H., & Sapers, G. M. (2000). Attachment and growth of salmonella chester on apple fruits and in vivo response of attached bacteria to sanitizer treatments. *Journal of Food Protection*, *63*(7), 876–883. https://doi.org/10.4315/0362-028x-63.7.876

Application of Optimized Sensible Heat Flux Calculation Combining Low-Pass Filtering and Wavelet Transform in the Study of Urban Thermal Environment in Jiangsu Province

Diwei Du[1*]

[1]Reading Academy, Nanjing University of Information Science & Technology, Nanjing, China

Abstract. The urban heat island effect is a common phenomenon during urbanization, characterized by significantly higher temperatures in urban areas compared to surrounding rural areas. Sensible heat flux (**H**) and temperature are important factors affecting the urban heat island effect. By optimizing the calculation of sensible heat flux, the urban thermal environment can be assessed more accurately, thereby guiding urban planning and green space layout to mitigate the heat island effect. This study proposes a method to optimize the calculation of sensible heat flux by combining low-pass filtering and wavelet transform. Low-pass filtering removes high-frequency noise, while wavelet transform extracts multi-scale fluctuation components, thereby improving the accuracy and stability of sensible heat flux calculations. Experimental results show that this method significantly enhances the precision of sensible heat flux calculations and reduces noise interference, providing reliable data for urban thermal environment research. This paper discusses in detail the application of this method in the study of the urban thermal environment in Jiangsu Province.

Key words: Urban Heat Island (UHI), Sensible Heat Flux, Low-pass Filter, Wavelet Transform, Urban Climate

1 Introduction

The Urban Heat Island Effect (**UHI**) refers to the phenomenon where urban areas exhibit significantly higher temperatures compared to the surrounding rural or suburban areas.

1.1. Overview of Urban Heat Island Effect

The Urban Heat Island Effect is primarily caused by human activities and urban construction, leading to changes in land use, the thermal properties of building materials, and the influence

*Corresponding author: 202183110014@nuist.edu.cn

of urban morphology on wind speed and humidity. The UHI effect not only affects energy consumption and air quality but also significantly impacts urban ecosystems, making it a critical area of study (Goldbach & Kuttler, 2013). Understanding these dynamics is essential for developing strategies to mitigate the UHI effect, particularly in rapidly urbanizing regions like Jiangsu Province (Li et al., 2019).

1.2 Importance of Sensible Heat Flux

Sensible heat flux (**H**) is a key indicator describing the heat exchange between the earth's surface and the atmosphere. Sensible heat flux transfers heat through turbulent motion, influencing the surface energy balance, boundary layer height, and local climate. Accurate measurement and calculation of sensible heat flux are crucial for understanding and simulating the urban thermal environment, predicting climate change, and formulating environmental policies. However, traditional methods for calculating sensible heat flux often face challenges due to the presence of high-frequency noise and low-frequency trends in the data, leading to inaccuracies and instability in the results (Nazari et al., 2012). Advanced data processing techniques such as wavelet transform and low-pass filtering have been introduced to overcome these challenges, offering more reliable data for urban heat environment studies (Gilles, 2013).

1.3 Limitations of Existing Methods

The calculation of sensible heat flux typically relies on meteorological data, such as temperature, wind speed, and humidity. However, these data often contain significant amounts of noise and trends that can affect the accuracy of the calculations. Traditional methods, like Fourier transform-based approaches, may not effectively remove these interferences, leading to unstable results with large errors. The wavelet transform provides a powerful tool for analyzing non-stationary signals by decomposing them into different frequency components, allowing for more precise denoising and trend removal (Hotra et al., 2022). This method, when combined with low-pass filtering, significantly enhances the accuracy and stability of sensible heat flux calculations.

1.4 Research Objectives and Methods

This paper proposes an optimized method for calculating sensible heat flux by combining low-pass filtering with wavelet transform. The objective is to improve the accuracy and stability of these calculations by effectively removing both high-frequency noise and low-frequency trends from the data. Low-pass filtering is used to retain the main signal components by eliminating high-frequency noise, while wavelet transform further refines the data by extracting fluctuation components through multi-scale analysis (Sang, 2013). This combined approach is expected to provide more precise and stable calculations of sensible heat flux, which is crucial for accurately assessing the urban thermal environment.

1.5 Background of Urban Thermal Environment Research in Jiangsu Province

Jiangsu Province, located in eastern China, is an economically developed and densely populated region. The rapid pace of urbanization in Jiangsu has exacerbated the urban heat island effect, making it increasingly important to accurately assess the thermal environment across various cities in the province. By applying the optimized sensible heat flux calculation method discussed in this paper, more reliable data can be obtained, which will support urban planning and the design of green spaces aimed at reducing the heat island effect (Sun et al., 2017).

2 Data and Methods

2.1 Data Sources

The data for this study comes from two parts: The first part consists of observational data from Guandu Village, Wugang Town, Quanjiao County, Chuzhou City, Anhui Province, China (**Fig. 1**). The observational data includes measurements taken on June 13, 14, and 15, 2019, with a sampling interval of 30 minutes. The variables measured include horizontal wind speed (**U**), vertical wind speed (**W**), temperature (**T**), and water vapor content (H_2O).

The second source comprises ERA5 reanalysis data, which provides large-scale and long-term data on sensible heat flux, temperature, and vertical velocity. This reanalysis data is particularly useful for analyzing the urban heat island effect in Jiangsu Province (Metzger et al., 2013).

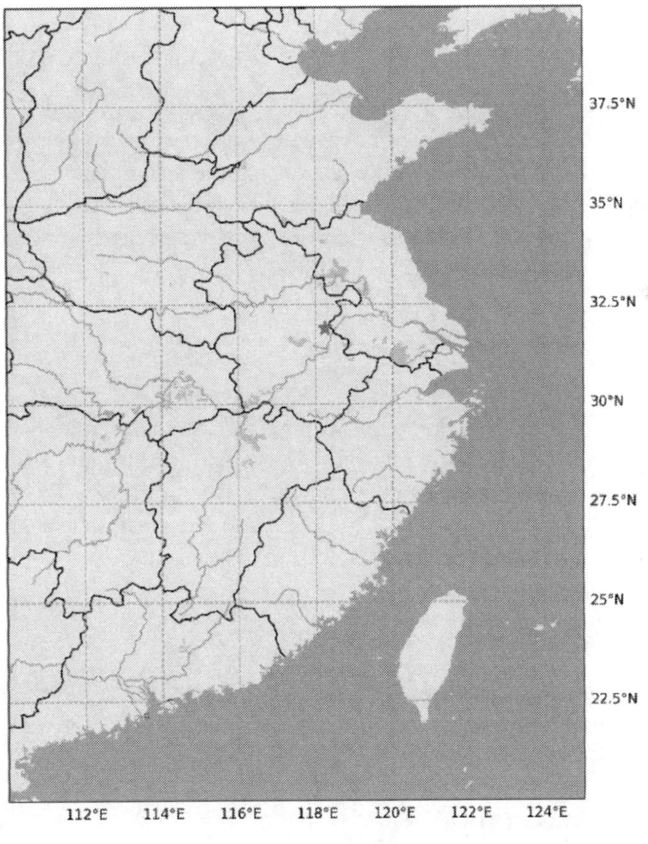

Fig. 1. The place where the observational data be measured (the red star)

2.2 Data Processing Steps

2.2.1 Data Selection

To ensure consistency and reliability in the analysis, observational data from the same season (summer) was selected. This choice is based on the fact that radiative and convective activities are more intense during summer, which helps to better highlight the characteristics of the urban heat island effect (Li et al., 2019). Data from June 13, 14, and 15, 2019, was extracted from the experimental dataset. Additionally, meteorological data from the summer of 2023, provided by the ERA5 reanalysis dataset, was selected for analysis.

2.2.2 Mean Calculation

The extracted data was averaged to remove constant components, which helps in understanding the overall trend and provides a foundation for subsequent detrending and turbulent fluctuation calculations. This step simplifies the data, making it easier to identify underlying trends and fluctuations that are critical for accurate calculation of sensible heat flux (Gilles, 2013).

2.2.3 Coordinate Transformation

To facilitate analysis, the horizontal wind speed components (**U** and **V**) were transformed to align with the principal axis. This transformation was performed using the principal component direction of the wind speed, converting the horizontal wind speed components into a coordinate system consistent with the main wind direction. This process simplifies the calculation of turbulent fluctuations, making the data more manageable and the results more accurate (Donateo et al., 2017).

2.2.4 Detrending and Mean Removal

The data was detrended to eliminate low-frequency trend components. This involved fitting a polynomial to each variable and removing the trend term, followed by the removal of the mean to obtain turbulent fluctuation components (u', v', w', t', q'). These fluctuation components reflect the contribution of turbulence to the sensible heat flux, allowing for a more accurate calculation (Metzger et al., 2013).

2.2.5 Covariance Calculation

We calculated the covariance between turbulent fluctuations, especially between vertical wind speed (w') and temperature (t') fluctuations. The formula for calculating sensible heat flux (H) is as follows:

$$H = C_P \rho \overline{w'T'} \qquad (1)$$

where:
- ρ is the air density (kg/m³),
- C_P is the specific heat capacity at constant pressure (*J/kg/K*),
- $w'T'$ is the covariance of vertical wind speed and temperature fluctuations (K·m/s).

2.3 Sensible Heat Flux Calculation Method Combining Low-Pass Filtering and Wavelet Transform

2.3.1 Low-Pass Filtering

Low-pass filtering is a fundamental technique used to remove high-frequency noise from signals while retaining the main signal components. For this study, a Butterworth low-pass filter was chosen due to its smooth frequency response characteristics. By setting the filter order to 5, a balance between filtering effect and computational complexity was achieved. The cutoff frequency, determined based on the data sampling frequency of 10 Hz, was set at 0.1 Hz. This configuration ensures that signal components below 0.1 Hz are preserved, effectively removing high-frequency noise while maintaining the integrity of the primary signal (Chen et al., 2010). The simplicity and effectiveness of low-pass filters make them particularly valuable in improving the signal-to-noise ratio, which is essential for precise sensible heat flux calculations.

However, while low-pass filtering is effective in reducing high-frequency noise, it may not sufficiently address multi-scale noise inherent in complex meteorological data. This limitation is addressed by combining low-pass filtering with wavelet transform, which is particularly adept at handling such challenges (Hotra et al., 2022).

2.3.2 Wavelet Transform Denoising

Wavelet transform is a powerful tool for time-frequency analysis, particularly suited for processing non-stationary signals. Unlike traditional Fourier-based methods, wavelet transform decomposes a signal into various scales or frequency components, allowing for the extraction of both detailed and approximated signal characteristics. This multi-level decomposition is crucial for capturing the inherent multi-scale features of the signal, which are essential for accurate sensible heat flux calculations (Pérez-Priego et al., 2017).

In this study, the Daubechies wavelet (db4) was selected as the basis function due to its excellent time-frequency localization properties. The signal was decomposed into different scales, with the decomposition level set to 4 to ensure sufficient detail extraction and noise removal. This approach allows the wavelet transform to effectively handle the complexities of environmental data, making it an ideal choice for enhancing the accuracy of sensible heat flux calculations (Gilles, 2013).

After decomposition, soft thresholding was applied to the detail coefficients, a critical step in the denoising process. The threshold was determined based on the noise level in the signal, utilizing the median absolute deviation (**MAD**) estimation method. By applying soft thresholding, noise is reduced while the essential components of the signal are retained. This not only improves the signal quality but also enhances the stability of the final sensible heat flux calculations (Nazari et al., 2012).

Finally, the denoised detail coefficients and the unprocessed approximation coefficients were reconstructed to obtain the final denoised signal through inverse wavelet transform. This process effectively combines the decomposed signal components back into a single, noise-reduced signal, ready for further analysis. The combination of low-pass filtering and wavelet transform thus provides a robust method for improving the accuracy and stability of sensible heat flux measurements, especially in the context of urban heat island studies (Pérez-Priego et al., 2017).

3 Experimental Results

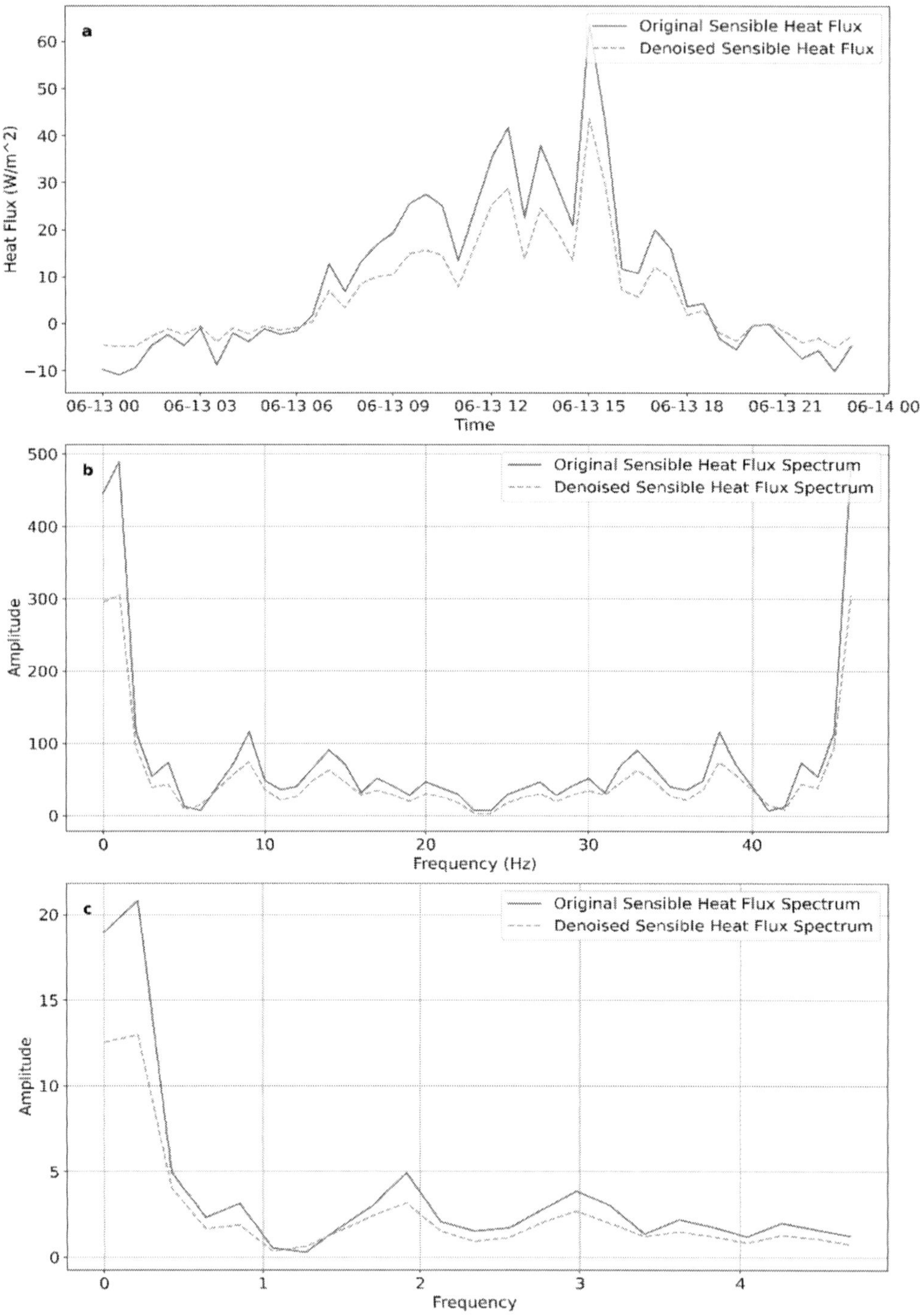

Fig. 2. The sensible heat flux and spectrum analysis on June 13, 2019. The top graph (a) compares the original sensible heat flux with the denoised flux. The middle graph (b) presents the spectra of the original and denoised fluxes. The bottom graph (c) shows the individual sensible heat flux spectrum.

https://doi.org/10.1051/e3sconf/202562003001

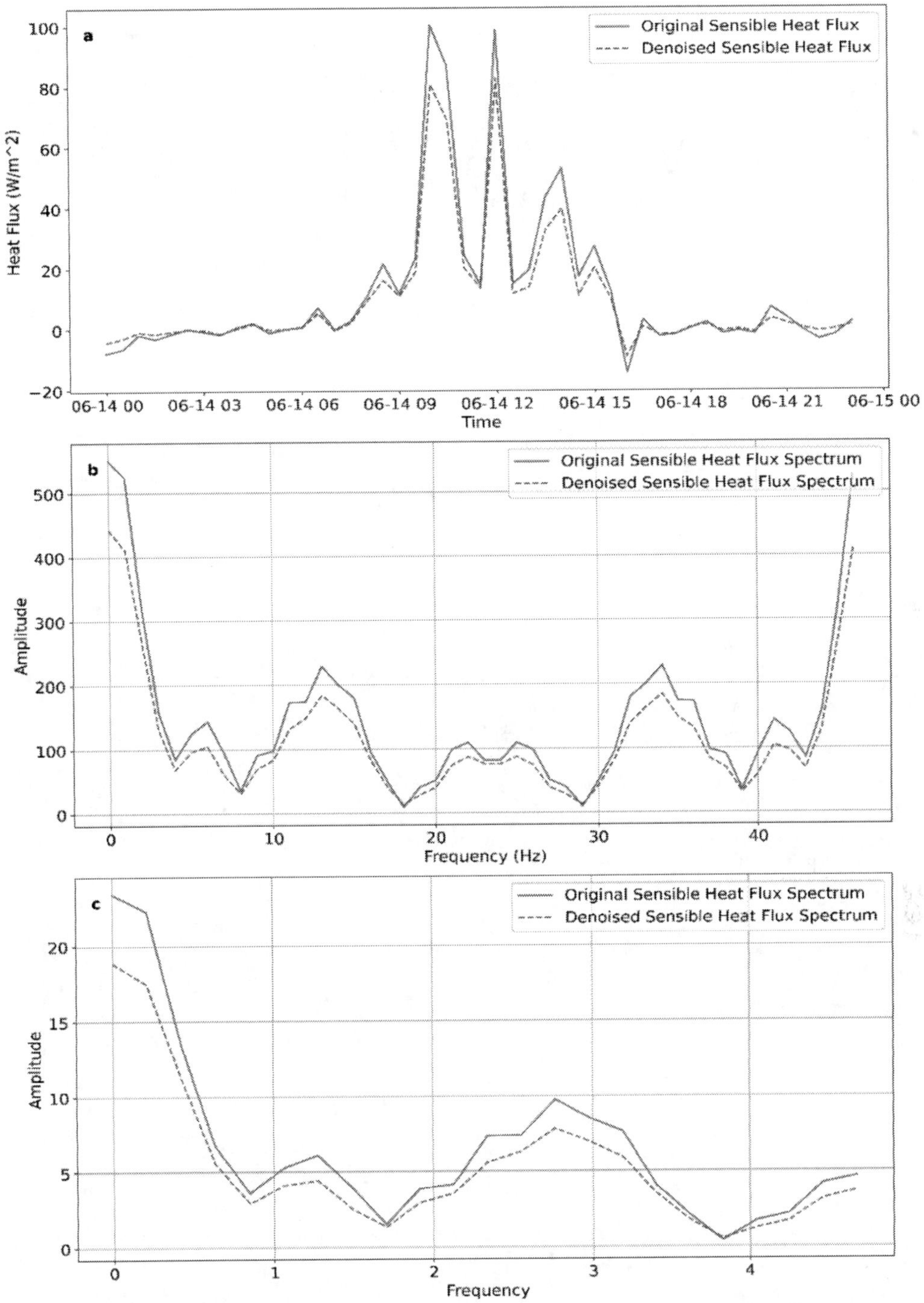

Fig. 3. The sensible heat flux and spectrum analysis on June 14, 2019. The top graph (a) compares the original sensible heat flux with the denoised flux. The middle graph (b) presents the spectra of the original and denoised fluxes. The bottom graph (c) shows the individual sensible heat flux spectrum.

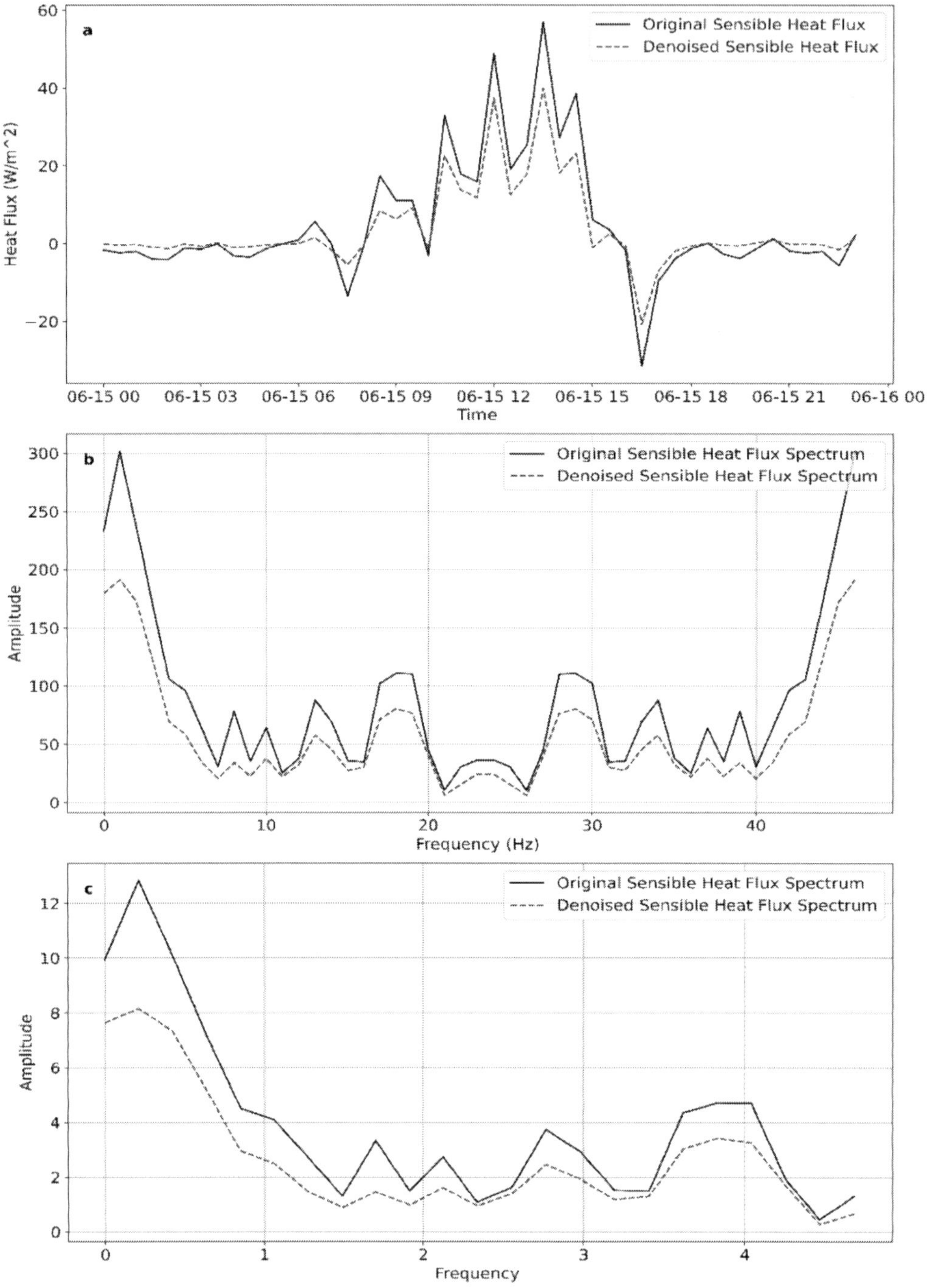

Fig. 4. The sensible heat flux and spectrum analysis on June 15, 2019. The top graph (a) compares the original sensible heat flux with the denoised flux. The middle graph (b) presents the spectra of the original and denoised fluxes. The bottom graph (c) shows the individual sensible heat flux spectrum.

Specific implementation results are made based on the observational data and the comparison of Sensible Heat Flux Before and After Denoising (**Fig. 2, Fig. 3, Fig. 4**). Through comparison of the original sensible heat flux and the denoised sensible heat flux over three days (June 13-15, 2019). The analysis of sensible heat flux and its spectrum for three consecutive days revealed the effectiveness of the denoising method using low-pass filtering and wavelet transform. On June 13, the original sensible heat flux, represented by a blue solid line, shows significant fluctuations, whereas the denoised flux, represented by an orange dashed line, is smoother with less fluctuation. The sensible heat flux spectrum comparison indicates that the high-frequency noise is effectively removed, as seen in the reduced amplitude of the high-frequency part of the denoised spectrum. This trend is consistent for the next two days. The denoised flux curves for these days are also smoother and less fluctuating, with notable stability in the flux data, particularly during peak periods. The spectrum analysis for these days confirms the retention of main signal components in the low-frequency range and effective high-frequency noise reduction. Overall, the denoising method significantly enhances the accuracy of sensible heat flux calculations, providing a clearer and more reliable representation of actual conditions.

By comparing the sensible heat flux and spectrum analysis over three days, it is evident that the denoising method effectively reduces the impact of high-frequency noise, resulting in a smoother sensible heat flux curve that more accurately reflects actual heat transfer conditions. The denoised spectrum retains the main signal components in the low-frequency range while significantly reducing high-frequency components, proving the effectiveness of low-pass filtering and wavelet transform. This optimized calculation method significantly improves the accuracy of sensible heat flux calculations, providing more reliable data support for urban heat environment research. The original sensible heat flux fluctuates significantly during the day, while the denoised sensible heat flux is relatively stable, more accurately reflecting actual heat transfer conditions. Low-pass filtering effectively removes high-frequency noise but retains low-frequency trend components. Wavelet transform further removes multi-scale noise, making the calculation results more accurate and stable. Wavelet transform can handle non-stationary signals, making it suitable for complex environments in practical applications.

From the sensible heat flux standard deviation analysis results from June 13 to 15, 2019 in table.1. It is shown that the denoising process significantly reduced the standard deviation of the sensible heat flux, indicating a significant reduction in data volatility. Specifically, the standard deviation of the original sensible heat flux on June 13 was 16.90 W/m², while the denoised standard deviation decreased to 10.94 W/m²; the original standard deviation on June 14 was 25.17 W/m², and the denoised standard deviation decreased to 18.52 W/m²; the original standard deviation on June 15 was 15.60 W/m², and the denoised standard deviation decreased to 9.29 W/m². These results indicate that low-pass filtering and wavelet transform effectively reduced high-frequency noise and abnormal fluctuations in the data, improving the accuracy and stability of sensible heat flux calculations.

Here is function of stand deviation:

$$S = \sqrt{\frac{\sum(\bar{x}-x_i)^2}{n}} \tag{2}$$

Table 1. The stand derivation of the original and denoised sensible heat flux in three days

Date	STD of original H	STD of denoised H
June 13	16.89	10.93
June 14	25.17	18.5
June 15	15.6	9.28

4 Urban Heat Environment Research in Jiangsu Province

4.1 Application of Optimized Sensible Heat Flux Calculation

By applying the optimized calculation method, combining low-pass filtering and wavelet transform, this study assesses the urban heat environment across various cities in Jiangsu Province. The use of reanalysis data on sensible heat flux and temperature, processed through the methods discussed, has provided more accurate and stable measurements, which are essential for understanding the spatial distribution of the urban heat island effect in this region (Li et al., 2019).

One of the key findings is the lack of a significant linear relationship between sensible heat flux and temperature in Jiangsu Province. The correlation analysis conducted in this study (**Fig. 5**), represented by a coefficient of -0.32, indicates that sensible heat flux alone may not be a direct predictor of temperature variations. This insight is critical as it suggests that other factors, such as land use, urban morphology, and green space distribution, play significant roles in shaping the urban thermal environment
Here is the function for correction:

$$r = \frac{\sum_{i=1}^{n}(X_i-\bar{X})(Y_i-\bar{Y})}{\sqrt{\sum_{i=1}^{n}(X_i-\bar{X})^2 \sum_{i=1}^{n}(Y_i-\bar{Y})^2}} \tag{3}$$

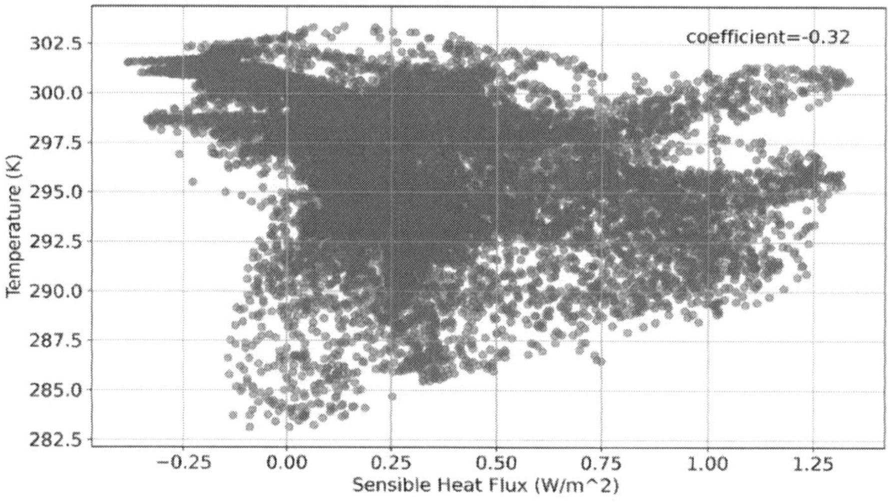

Fig. 5. The scatter between sensible heat flux and temperature with the value of coefficient.

 https://doi.org/10.1051/e3sconf/202562003001

4.2 Sensible Heat Flux Distribution Across Cities

The study further examines the distribution of sensible heat flux across different cities in Jiangsu Province (**Fig. 6, Fig. 7, Fig. 8, Table 2**). The analysis reveals that cities with higher levels of urbanization, such as Nanjing and Wuxi, exhibit higher sensible heat flux values, indicating a more pronounced urban heat island effect. This is likely due to the high density of buildings and industrial activities, which increase the heat load in these areas. Conversely, cities with more green spaces and water bodies, such as Suzhou and Lianyungang, show lower sensible heat flux values, reflecting their more effective natural cooling mechanisms (Sun et al., 2017).

These findings align with previous research that highlights the critical role of urban planning in mitigating the urban heat island effect. The presence of green spaces and water bodies not only helps to lower temperatures but also contributes to a more balanced and sustainable urban environment. As such, these elements should be prioritized in urban development strategies to reduce the adverse effects of urbanization on local climates (Goldbach & Kuttler, 2013).

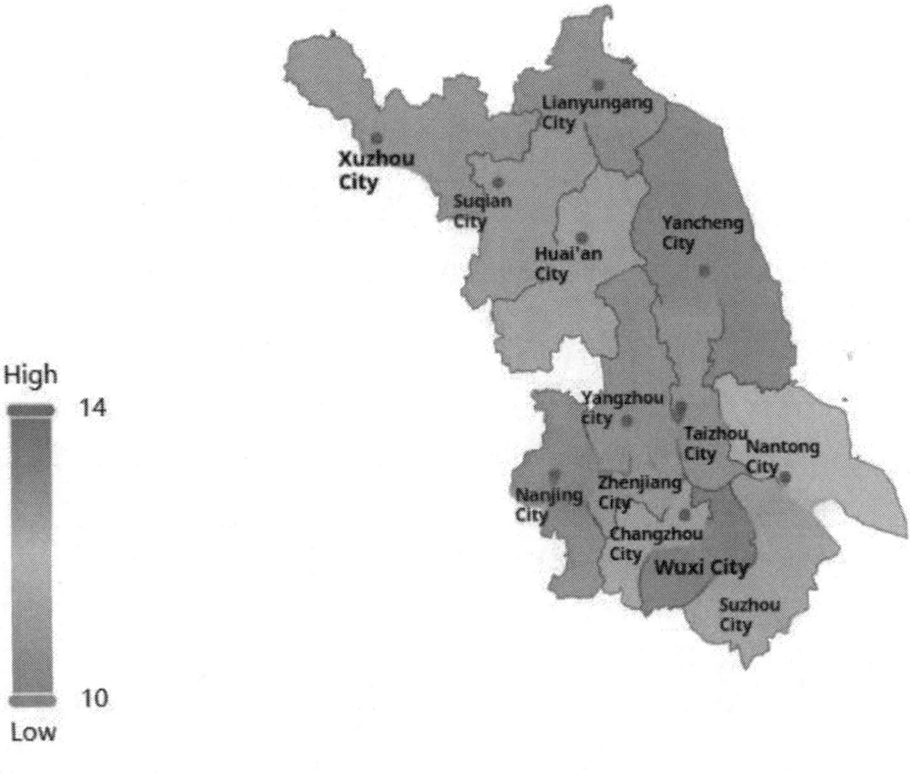

Fig. 6. The distribution of sensible heat flux in Jiangsu province.

Fig. 7. The surface temperature distribution in Jiangsu province.

Table 2. The sensible heat flux and surface temperature in cities in Jiangsu province

Cities	Sensible heat flux (W/M^2)	Surface temperature (K)
Nanjing	10.71	297.41
Wuxi	13.30	298.02
Xuzhou	10.79	298.68
Changzhou	11.52	297.71
Suzhou	11.40	297.51
Nantong	12.05	297.00
Lianyungang	10.73	297.62
Huai'an	11.49	297.34
Yancheng	10.46	295.83
Yangzhou	10.89	296.81
Zhenjiang	11.26	297.15
Taizhou	10.96	296.63
Suqian	11.24	297.95

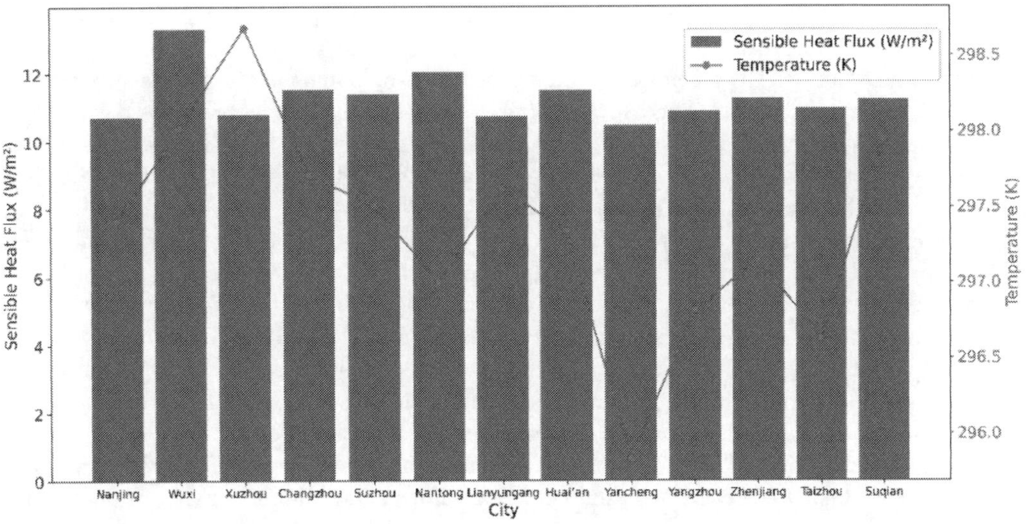

Fig. 8. The temperature and sensible heat flux of different cities

5 Analyses and suggestions according to cities in Jiangsu province

Based on the analysis of cities in Jiangsu Province, it is evident that the urban heat island effect varies widely depending on the level of urbanization, industrial activity, and the presence of natural features like green spaces and water bodies. Nanjing, the capital of Jiangsu, suffers from a significant urban heat island effect due to its dense population, frequent industrial activities, and extensive urban development. To mitigate this, it is essential to increase urban green spaces, optimize ventilation corridors, and promote the use of energy-efficient building materials.

Similarly, Wuxi, a major industrial city, experiences high sensible heat flux and elevated temperatures as a result of dense industrial emissions and infrastructure. To reduce the heat island effect, it is recommended to strengthen industrial emission control, increase urban green spaces, and implement green building standards. Xuzhou, another heavily industrialized city, faces similar challenges with high sensible heat flux due to its rich mineral resources and dense development. Expanding urban green spaces and controlling industrial emissions in these cities would be crucial steps in mitigating the heat island effect.

In contrast, Suzhou benefits from well-maintained green spaces and water bodies, which help moderate the urban heat island effect, even in the face of significant urbanization. Suzhou's sensible heat flux and temperature are relatively moderate, thanks to its balanced urban layout. Continuing to prioritize the construction of green spaces and water bodies, while optimizing the urban layout, will help the city further reduce the impact of urban heat. Changzhou presents a balanced scenario, with moderate sensible heat flux and temperature due to a mix of industrial activities and large green spaces. Increasing the number of green spaces and improving industrial emission controls would help reduce the heat island effect in the city. Yangzhou, a tourist city, emphasizes ecological protection and cultural heritage, resulting in a relatively weak heat island effect. The city can maintain this balance by further enhancing green spaces, reducing vehicle emissions, and optimizing traffic layouts.

Coastal cities like Lianyungang and Yancheng benefit from their proximity to water, which naturally mitigates the heat island effect. Lianyungang has a mild climate with low sensible heat flux, and its green spaces and water bodies help maintain this balance. Similarly, Yancheng's mix of industrial and agricultural activities, along with large green spaces, keeps its sensible heat flux and temperature moderate. Maintaining and expanding these natural landscapes, along with optimizing industrial layouts, will be key to preventing the urban heat island effect from intensifying in these cities.

In Zhenjiang, the favourable geographical location and balanced urban construction between industrial zones and green areas help moderate sensible heat flux. To further alleviate the urban heat island effect, the city should focus on optimizing its industrial layout and enhancing its green spaces. Taizhou, which has experienced rapid urbanization while paying attention to environmental protection, demonstrates a moderate heat island effect. Continued focus on increasing green spaces and reducing emissions from industrial and traffic sources will help keep the urban thermal environment in check.

Huai'an, with its balanced industrial, agricultural, and commercial activities, faces moderate heat island effects due to industrial emissions and building density. Strengthening industrial emission controls and expanding urban green spaces will help reduce the heat island effect. Lastly, Suqian, an emerging city, shows moderate sensible heat flux and temperature due to rapid urbanization. Strategic urban planning that focuses on green space development, emission reduction, and the optimization of ventilation corridors will be key to managing the urban heat island effect as the city continues to grow.

In summary, cities like Nanjing, Wuxi, and Xuzhou exhibit higher sensible heat flux due to dense urbanization and industrial activities, requiring interventions such as expanding green spaces and controlling emissions. In contrast, cities like Suzhou, Lianyungang, and Yancheng benefit from natural landscapes that mitigate the heat island effect, and their focus should be on maintaining this balance. Cities with moderate heat island effects, such as Changzhou, Yangzhou, Zhenjiang, Taizhou, Huai'an, and Suqian, should continue to prioritize green space development and sustainable urban planning to prevent further intensification of urban heat.

6 Conclusion

This study highlights the importance of advanced data processing techniques, specifically the combination of low-pass filtering and wavelet transform, in improving the accuracy of sensible heat flux calculations. Traditional methods often struggle with high-frequency noise and low-frequency trends, reducing accuracy. The proposed method effectively addresses these issues, enhancing stability and precision in assessing urban heat environments, particularly in regions like Jiangsu Province.

Key findings reveal that sensible heat flux alone does not have a simple linear correlation with temperature in urban areas, suggesting that other factors, such as land use, urban design, and green space distribution, significantly impact the urban thermal environment. The spatial analysis of heat flux across cities in Jiangsu shows that highly urbanized areas, such as Nanjing and Wuxi, exhibit higher flux values, while cities with more green spaces, such as Suzhou, show lower values. This underscores the role of urban planning in mitigating the urban heat island effect.

Recommendations include increasing green spaces and water bodies, optimizing urban layouts for better ventilation, and promoting energy-efficient building materials. Additionally, addressing industrial emissions is critical, as they significantly contribute to urban heat. This method offers a practical framework for improving urban heat management and can guide city planners and policymakers in developing strategies to mitigate the adverse effects of urbanization.

In conclusion, the optimized calculation method provides valuable insights into urban heat dynamics, helping cities reduce the urban heat island effect and promote sustainable urban growth. Continuous monitoring and data-driven planning are essential for adapting these strategies to the challenges of ongoing urbanization and climate change.

References

1. Chen, H.-P., Liao, H.-J., Huang, C.-M., & Yu, S.-N. (2010). Improving liquid chromatography-tandem mass spectrometry determinations by modifying noise frequency spectrum between two consecutive wavelet-based low-pass filtering procedures. Journal of Chromatography A, 1217(17), 2804-2811.
2. Donateo, A., Cava, D., & Contini, D. (2017). A case study of the performance of different detrending methods in turbulent-flux estimation. Boundary-Layer Meteorology, 164(1), 19-37.
3. Gilles, J. (2013). Empirical Wavelet Transform. IEEE Transactions on Signal Processing, 61(16), 3999-4010.
4. Goldbach, A., & Kuttler, W. (2013). Quantification of turbulent heat fluxes for adaptation strategies within urban planning. International Journal of Climatology, 33(10), 2340-2348.
5. Hotra, O., Kovtun, S., Dekusha, O., Grądz, Ż., Babak, V., & Styczeń, J. (2022). Analysis of Low-Density Heat Flux Data by the Wavelet Method. Energies.
6. Li, H., Zhou, Y., Wang, X., Zhou, X., Zhang, H., & Sodoudi, S. (2019). Quantifying urban heat island intensity and its physical mechanism using WRF/UCM. The Science of the Total Environment, 650 Pt 2, 3110-3119
7. Metzger, S., Junkermann, W., Mauder, M., Butterbach-Bahl, K., Widemann, B. T., Neidl, F., Schäfer, K., Wieneke, S., Zheng, X., Schmid, H., & Foken, T. (2013). Spatially explicit regionalization of airborne flux measurements using environmental response functions. Biogeosciences, 10(4), 2193-2217.
8. Nazari, M., Farahani, S. D., & Kowsary, F. (2012). Comparison of the Mollification and Wavelet Prefiltering of Temperature Data in an Ill-Posed Inverse Heat Conduction Problem, Application: Nonthermal Equilibrium Porous Medium. Heat Transfer Engineering, 33(8), 704-711
9. Onderka, M., Banzhaf, S., Scheytt, T., & Krein, A. (2013). Seepage velocities derived from thermal records using wavelet analysis. Journal of Hydrology, 479, 64-74.
10. Pérez-Priego, Ó., El-Madany, T., Migliavacca, M., Kowalski, A., Jung, M., Carrara, A., Kolle, O., Martín, M., Pacheco-Labrador, J., Moreno, G., & Reichstein, M. (2017). Evaluation of eddy covariance latent heat fluxes with independent lysimeter and sapflow estimates in a Mediterranean savannah ecosystem. Agricultural and Forest Meteorology, 236, 87-99.
11. Sang, Y. (2013). A review on the applications of wavelet transform in hydrology time series analysis. Atmospheric Research, 122, 8-15
12. Sun, T., Wang, Z., Oechel, W., & Grimmond, S. (2017). The Analytical Objective Hysteresis Model (AnOHM v1.0): methodology to determine bulk storage heat flux coefficients. Geoscientific Model Development, 10(8), 2875-2890.

Bio Thermal Panels Applied to the Comfort of Rural Homes in the District of Sincos in 2024

Vladimir Montoya Torres[1*], *and Jhomara Rojas Tacza* [2]

[1] Universidad de Huánuco, Huánuco 10001, Perú
[2] Universidad Continental, Huancayo 12002, Peru

Abstract. Within the context of the Sincos District in its high Andean areas, there is the Aramachay Community that suffers from the intense cold. The study proposes better thermal comfort in the face of low temperatures inside the homes through a biothermal insulator made of materials from the area, during the period 2023. The research is quasi-experimental since the results that are measured will be described. The research method consists of making biothermal panels with materials of natural origin and seeking thermal comfort in rural homes. The first step is to obtain data on temperature, humidity and wind with instruments using the method without insulation and with insulation. The bio thermal panel is made with local materials, two panels are made with different materials and the different results are compared, this helps us make a decision as to which panel is more viable for application in thermal insulation at the level of the covers of rural homes. The results show that the first test of Ichu Fiber, Barley Straw and Gypsum obtained the lowest temperature at 0.7 °C with humidity of 61.6%Rh, the second test of Ichu Fiber, Barley Straw and Clay the lowest temperature low is 1.1 °C with humidity of 63.9%Rh, which determines that both bio thermal panels are thermal insulators, 1.1 °C with humidity of 63.9%Rh, which determines that both bio thermal panels are thermal insulators, this means that the prototype has a temperature improvement of 1.7 °C with a biothermal panel, gaining heat inside rural homes, gaining heat inside rural homes. This constitutes a possibility of how we can provide better thermal comfort with an alternative based on sustainable development, because it generates social, environmental and economic impacts for the population.

Keywords: Bio Thermal Panels , Thermal Comfort , Rural Housing.

1 Introduction

In the department of Junín there are four very marked ecological levels, of which those in the high mountains have temperatures that drop significantly, registering up to periods of four days followed by frost, mainly in the months of June, July and August, the climate is rainy and cold with dry autumn and winter, where the maxi-mum temperature ranges throughout the year between 14°C in March and 16°C in November and the minimum temperature, between 2°C and 3°C in the summer months and -6°C to - 4°C in the winter months [1],

* Corresponding author: vladimir.montoya@udh.edu.pe

descending between 05:30 and 07:30 in the morning. Precipitation in the Mantaro Valley occurs during the months of January, February and March; February is one of the rainiest months with 50% of annual precipitation which produces a higher percentage of humidity and consequently intense cold with an average minimum temperature -2 °C and 83% humidity which is evidenced by the cold thermal sensation in the months of January to March, with the month of July being the lowest month by 48% [2].

Within the Mantaro Valley is the District of Sincos, one of the districts of the province of Jauja, located at 3330 meters above sea level, the temperature varies from 5 °C to 20 °C on the coldest days, dropping by 2 °C; In the high Andean population centers of the Sincos district, there are periods of frost reaching below -8 °C, affecting the towns of Cruz Pampa, Sallahuachac , San Juan de Miraflores, Liglish Pampa, Chalhuas , Huaripata, Aramachay among others. who are located at more than 3900 meters [3] above sea level under these weather conditions, the most sensitive to the intense cold are the elderly, children and infants residing in the town center of Aramachay.

Rural homes have predominant characteristics such as corrugated iron roofs, ado-be walls, stone with mud and dirt floors; they do [4] not have the architectural principles, materiality and comfort parameters [5], with a deficiency of thermal comfort in the interior space and this affects directly on people's quality of life. Most roofs are replaced due to the ease of installation and low cost with galvanized corrugated iron, this is a material that does not thermally insulate from the outside, allowing cold wind to enter the interior. The low temperature in the homes is a product of humidity, this is a result of not having a finished floor and they are sitting on the natural soil [6]. The main problem of the precariousness of housing is the roofs using materials such as corrugated iron and tiles, in this sense the intervention using biothermal panels is analyzed, based on preserving the warm temperature by functioning as a thermal insula-tor against cold temperatures [7], the panels will be composed of sustainable natural fibers that do not affect people and do not contribute to the pollution of the planet [8], providing thermal comfort in homes.

Table 1. Materials, technical specifications.

Materials	Technical specifications	
Stipa Ichu Fiber It is a semi-dry pasture native to more than 3,400 meters above sea level used as fodder for animals As can be seen in Figure 01. [9]	Condition: semi dry	Length:60 – 100cm
	Thermal conductivity: 0.007 W/mK	Moisture in dry fiber: 15%
	Thermal insulator	Cost: recycled
Barley Straw They are residual dry stems that are obtained once the barley is harvested or the barley is threshed. It is more flexible compared to other types of straw such as wheat or oats, its texture is softer, and it has thermal and acoustic insulation qualities, This material can be visualized in Figure 2 . [10]	Condition: Dry	Length: 60 to 100 cm high
	Thermal conductivity: 0.53 to 0.04 W/ mK	Humidity: 14.2%
	Thermal and acoustic insulation	Cost: recycled
Plaster	Condition: Kneaded with water	Quantity: 8kg

It is a mineral made up of calcium sulfate, it is a binding material that serves to join other materials, it has thermal and acoustic insulating properties for walls and ceilings, low weight, low cost, this material can be visualized in Figure 3. [11]	Thermal conductivity: 00.25 to 0.50 W/ mK	Humidity: 1.2%
	Thermal and acoustic insulation	Cost: 1 dolar x 1 bag
Clay Clay is a decomposed sedimentary rock made up of hydrated aluminum silicates preceding the decomposition of rocks containing feldspar. When it is pure it presents red colors, orange to white when it is pure. It is of very light ceramic origin with a highly porous structure, it is a thermal and acoustic insulating material, this material can be visualized in Figure 4. [11]	Condition: kneaded with water	Feature: retains up to 80% internal heat
	Thermal conductivity: 0.1 – 0.18W/ mK	Thermal Resistance: 0.121 - 0.714 (m2-k/W) Humidity: 12%
	Thermal and acoustic insulation	Cost: Obtained from land at no cost

Fig. 1. Barley Straw material.

Fig. 2. Material Stipa Ichu fiber.

Fig. 3. Material Clay.

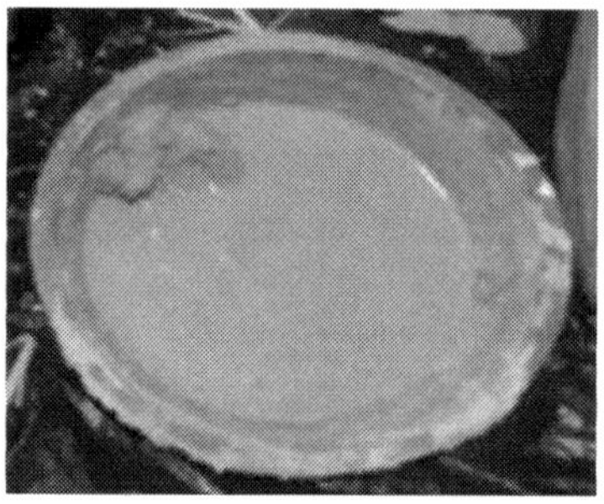

Fig. 4. Material Plaster.

Table 2. Summary table of Instruments, technical specifications and photographs.

Instruments	Technical specifications	
Thermometer + Hygrometer Calibrated professional instrument, As can be seen in Figure 5.	Brand: Benetech	Model: GM 1361
	Temperature range: -10~50°C (-20~50 K type)	Selection between °C and °F
Anenometer Calibrated professional instrument, As can be seen in Figure 6 .	Brand: UNI-T Model UT 363	Wind Accuracy : ± (5% reading + 0.5)

 https://doi.org/10.1051/e3sconf/202562003002

Insulation Molds They are molds to make insulators, suitable for the shape of calamine, As can be seen in Figure 7.	Measurement: 0.46 x 0.46 x 0.05 cm Material: phenolic plywood	Base: transparent calamine
Prototype A prototype measuring 50 x 50 x 50 cm is made of phenolic plywood material. Prototype similar to a house with a corrugated board roof, suitable for placing insulation, As can be seen in Figure 8.	Extent: Height 1: 50 cm Height 2: 60 cm Width: 50 cm Length: 50cm	Cover: calamine
	Material: phenolic plywood	

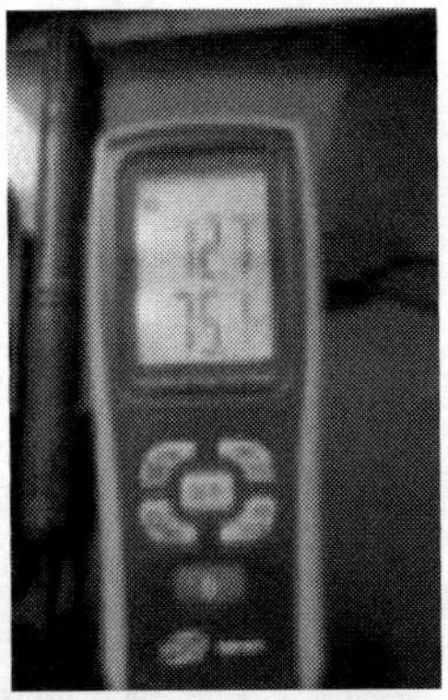

Fig. 5. Thermometer Hygrometer instrument.

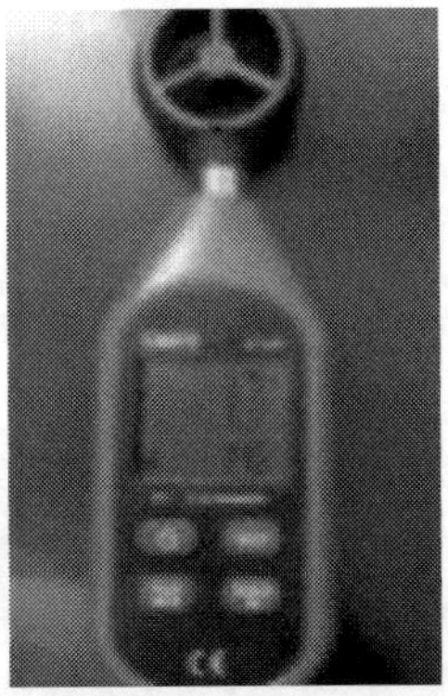

Fig. 6. Anenometer Instrument.

Fig. 7. Mold for the Insulator.

Fig. 8. Prototype for Recording Thermal Measurements.

2 Method

This research is of a descriptive type and at a correlational descriptive level. The design is quasi-experimental, since the results achieved when measuring will be described. The variables are bio thermal panel materials of natural origin that allow homes to be insulated from extreme temperatures and thermal comfort is a neutral sensation of the person in an environment. It is investigated in the high Andean areas of the Sincos district in the Province of Jauja, Department of Junín, the community of Aramachay is located at 3688 meters above sea level [12] with an average temperature of 15° at night, the lowest temperature reaching an average of 2°C. It also drops to -12 °C in the early morning [13] and the average annual precipitation is 1581 mm, the average humidity is 69%. [14]. In order to achieve an adequate temperature measurement, work was done using a summary table of instruments, technical specifications and photographs, these can be seen in table 2.

 https://doi.org/10.1051/e3sconf/202562003002

The community of Aramachay is located in the center of the foot of four hills, in the middle of the place there is a meadow or lagoon as can be seen in Fig. 9 of Google Earth, its main economic activity is agriculture, being one of its main crops. Barley is one of the products that is most planted, obtaining the Barley straw that serves as food for animals and material for construction in adobes and rammed earth. Likewise, Ichu, its scientific name is Stipa Ichu, is a widely used plant. As fodder for the mammals that live in the Andes, it is also used as construction material in the production of adobe and roofing of huts. It [15] grows in large quantities in the community of Aramachay. These solid wastes are generated by agriculture in large quantities of fibers. of straw, due to the large volume of straw produced by these cereals such as barley, wheat, oats, etc.; It is used as food forage for animals, however, since it is not used, most of it is discarded or in large quantities, the practice of burning stubble is used to facilitate the preparation of the land for the next sowing [16].

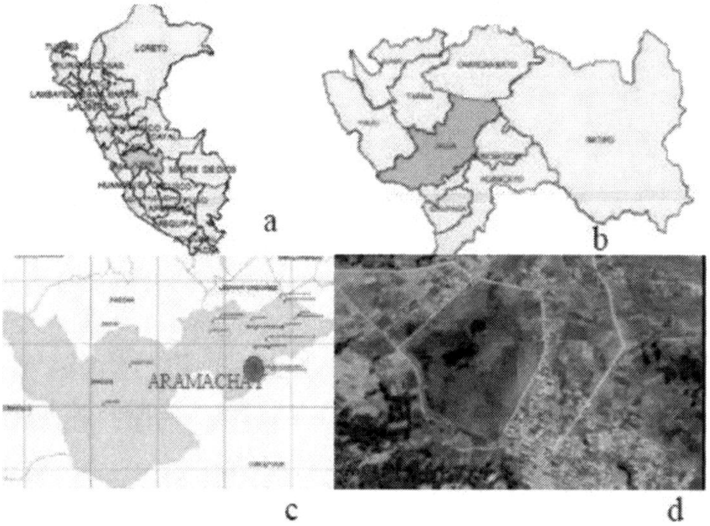

Fig. 9. (a) Location map of Junin, (b) Location map of Jauja, (c) Location map of Sincos, (d) Location map of Aramachay.

To achieve the objective of this research, a model similar to a house made of phenolic plywood material with a corrugated iron roof was made. This model simulates a house which helps us obtain data with and without the insulation. For the construction of the insulation, a mold is used considering the waves of the corrugated board, achieving minimal air leaks, cold and heat loss inside the rural home, eliminating thermal bridges. A thermal bridge is a place of a structure with significantly worse thermal properties than the rest of the structure [17]. Ichu and Barley straw have the characteristic of thermal insulators; to achieve the compaction of these materials, Plaster and Clay are used independently, both are low-cost and easily accessible materials as described in Table 1 materials. Clay is found in the earth and is a thermal insulating material.

2.1 Panel layout:

Table 3. Steps for panel design, description and photographs.

01	**Mold Design** The shape of the calamine helps that there are no wind and cold leaks in the homes and that it adapts to the corrugated board coverage.

https://doi.org/10.1051/e3sconf/202562003002

02	**Material Selection** Once the size of the insulation is defined, the bio thermal panels are made. The materials were selected such as Ichu natural fibers and barley straw, the cleaning and selection of the natural fibers was carried out in a semi-dry condition, the Ichu was used in greater quantities as it is a thermal insulator [18].	Barley Straw Measurement: Cut from 8 to 10 cm. Condition: Semi dry. Ichu Measurement: Cut from 8 to 10 cm. Condition: Semi dry. Clay Measurement: 12 kilograms of pure clay. Condition: soak in water for 3 days to achieve a more durable and moldable one. Plaster Measurement: a 4 kilograms bag Condition: the dough properly mixed with water until a paste is obtained.
03	**Realization of Bio Thermal Panels** Having all the materials, we proceed to make the bio thermal panels, the materials are combined to obtain greater thermal comfort, These details can be seen in figure 10.	Plaster biothermal panel with Ichu and 5 cm Barley straw The Ichu is sprayed with water to mix it in a plaster container with water with an equal amount of both and make a homogeneous mixture. Having the mold of the panels, a light layer of Ichu is placed intertwined one on top of the other and the paste is immediately made. of Plaster, then a layer of Ichu was placed along with a little Barley straw. Finally, a layer of the mixture of Plaster with Ichu was placed, achieving the 5 cm that one wants to obtain for the panel. Clay biothermal panel with Ichu and 5cm Barley straw The same steps are carried out for Plaster panel, but it is replaced with Clay. Finally, let it dry in the sun after two days and proceed to unmold the panel.
04	**Model Design** For the investigation, a model similar to a house with a corrugated iron roof with a drop is developed, with a base size of 0.50 x 0.50 cm and a height of 0.60 and 0.50 cm, it is carried out to obtain the data that is required. The design of the model is suitable for placing the bio thermal panel.	
05	**Placement Of The Bio Thermal Panel And Obtaining Data** The bio thermal panel is placed, which is supported by a kind of ridge on the model, the part of the waves is placed on the part of the corrugated board and thus makes it fit well and does not filter the wind and cold into the interior of the model. Once the insulator and the thermometer are placed inside, the data per day for each insulator is taken. The idea of this element is that it allows the modulation of the ceiling with a resistance of 20 to 25 kg per square meter, These details can be seen in figure 11.	

Fig. 10. Composition of the bio panel that will be inserted in a thermal measurement prototype.

Fig 11. Prototype for the control of thermal measurements.

3 Results

Below is a description of the results obtained and tables showing the measurements made with and without the insulator in the model, considering the day, they were taken to record temperature, humidity and wind measurements.

In order to properly record these results, it was necessary to build the prototype, which describes its step-by-step implementation for the design of the panel and assembly, as well as its description. This can be seen in Table 3.

To obtain accurate results of temperature, humidity and wind in the Community of Aramachay, instruments such as Thermometer, Hygrometer and Anenometer are used. The first measurement at the study site is carried out with no type of insulation on May 23, 2023, recording the temperature. lower than -1 °C with humidity of 65.9 %Rh at 6:23 AM as seen in Fig. 12 shows the control of temperature, humidity and wind without insulation at the study site, it is recorded as the temperature data varies depending on the time and the wind while the temperature is quite low.

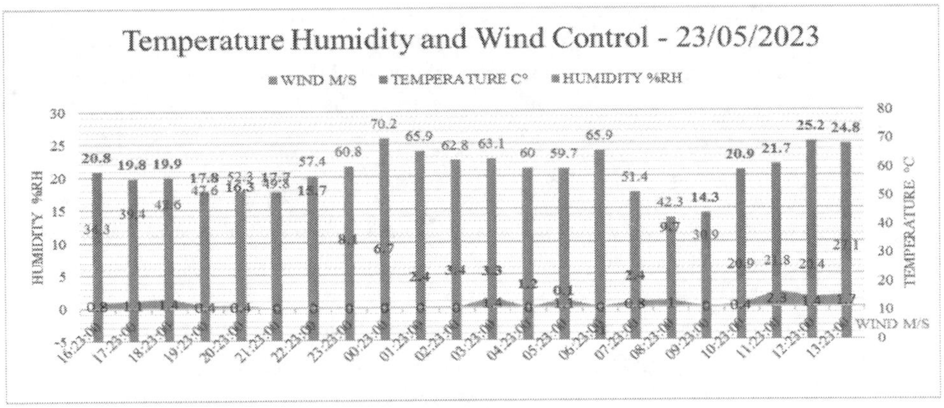

Fig. 12. Control of temperature, humidity and wind without insulation.

For the evaluation of the biothermal panels, two tests were carried out with different materials, evaluating thermal comfort, heat loss and gain of each of the materials used. The first test was carried out on June 16, 2023, the biothermal panel is made of Ichu Fiber, Barley Straw and Gypsum material to obtain the data needed, the biothermal panel is placed in the model as seen in the Fig. 13.

Fig. 13 shows how the temperature varies depending on the time, humidity and wind. Obtaining the lowest temperature at 6:29 am of 0.7 °C with humidity of 61.6%Rh, in addition the months of June , July and August are months of frost and the temperature drops to - 8 °C in the community of Aramachay [2]; demonstrating that the bio thermal panel of Ichu Fiber, Barley Straw and Plaster is a thermal insulator, achieving thermal comfort that gains and does not lose heat in the model.

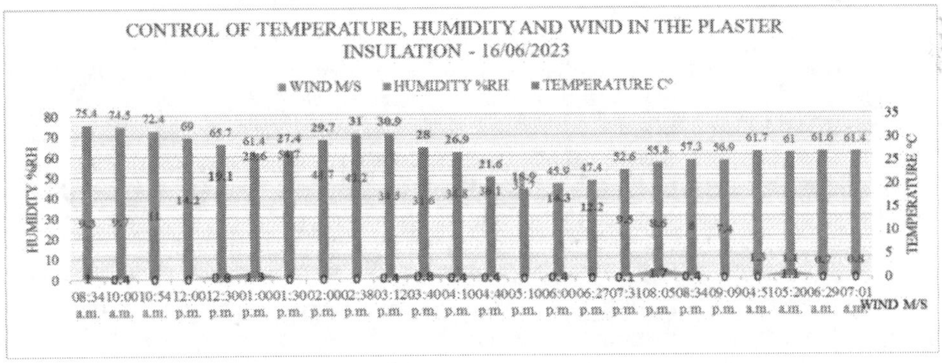

Fig. 13. Temperature, humidity and wind control with Ichu, Barley Straw and Plater material insulation.

The second test was carried out on June 17, 2023, the biothermal panel is made of Ichu Fiber, Barley Straw and Clay material. Fig. 13 shows how the temperature varies depending on the time, humidity and wind. Obtaining the lowest temperature at 6.51 am of 1.1 °C with humidity of 63.9%Rh, demonstrating that the bio thermal panel of Ichu Fiber, Barley Straw and Clay is a thermal insulator achieving thermal comfort that gains and does not loses heat in the model.

The research is framed in the need to improve habitability conditions in rural homes by taking advantage of innovative bioclimatic solutions. In this context, it is closely linked to the principles set out in the study Ancestral Earthen Construction Techniques Updated to the Needs of the People in the Central Andes of Peru, [20] which highlights the importance of recovering and updating vernacular construction techniques to respond to contemporary demands for comfort and energy efficiency.

Both investigations share the approach of integrating local materials and traditional knowledge with modern technologies to optimize thermal conditions in homes in Andean communities. While the first study explores the potential of biothermal panels as a sustainable solution, the second highlights the importance of training architecture students through community service-based learning experiences. Together, these investigations demonstrate that the combination of technological innovation and ancestral knowledge is key to the development of sustainable architectural solutions that improve the quality of life in rural contexts of the Peruvian Andes.

As can be seen in fig. 14 the use of Bio Thermal Panels Applied to the Comfort of Rural Housing in the District of Sincos in 2024, positively influence thermal insulation, these measures reflect how the use of this alternative reduces the exterior temperatures inside the environment where the panel is applied, the use of insulating materials has a notable influence[21], for this reason the experimental part is validated in a positive way as mentioned in the Moncloa 2017 research [6].

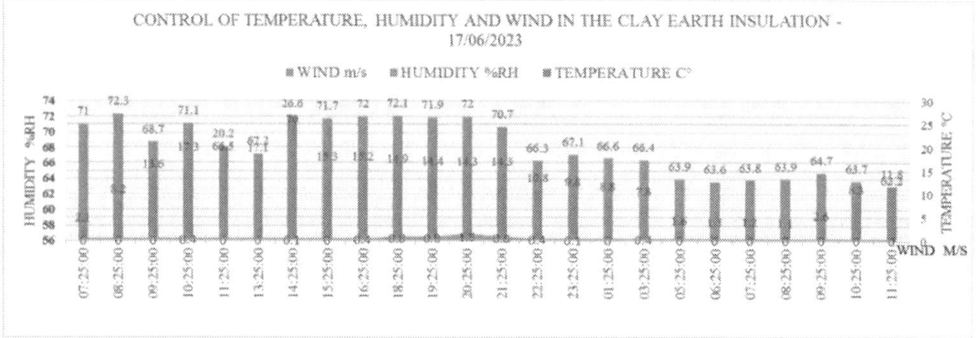

Fig. 14. Temperature, humidity and wind control with Ichu, Barley Straw and Clay material insulation.

4 Discussion

The following discussion is carried out: according to the tests carried out, it is confirmed that the materials used have the capacity for thermal conductivity and heat retention [7]; In his research he made panels with Barley and Wheat fibers, synthetic glue, sawdust and geogrids (0.60 x 0.50 x 0.04 cm), managing to reduce the heat flow in a home by 26%. However, the panel concluded that The panel has a cost of S/ 6.05, the proposal manages to reduce the reduction of polluting gases such as CO_2. At the same time, the investigation [19] reached the conclusion that the mixture of materials of 80% plaster and 20% sheep wool fiber generated a reduction of extreme temperatures inside the homes through panels or surface coverings, achieving the thermal comfort inside homes. The results of both investigations had a temperature differentiation according to the materials used and with a different type of mixture each.

In this article, natural materials are used, making biothermal panels, unlike previous research, these are panels for surface coverage placed as a type of ceiling, managing to retain heat, producing thermal comfort in the face of low temperatures, according to the tests carried out by the company. proposal manages to retain heat up to approximately 7 C° in both biothermal panels, biothermal panels have the capacity to be thermal insulators and more environmentally friendly, reducing the carbon footprint.

This study is closely related to the publication Lightened Blocks of Plant Fibers and Clay Applied in Construction Systems for Housing in High Andean Areas, which investigates the application of natural materials with low environmental impact in the construction of efficient homes in high Andean areas.

Both studies highlight the importance of using local materials, such as plant fibers, clay and biothermal elements, to optimize thermal insulation and reduce energy consumption in rural buildings. While biothermal panels seek to regulate the interior temperature of homes in a sustainable way, lightened blocks of plant fiber and clay propose a structural solution with improved thermal properties. Together, this research contributes to the development of resilient and sustainable construction systems, which combine technological innovation with traditional knowledge to improve the quality of life in Peruvian high Andean communities. [21]

The research is aimed at optimizing thermal comfort in rural homes through the use of biothermal panels, a sustainable strategy based on bioclimatic principles. This study finds a significant link with the publication Revaluation of Vernacular Architecture in a Hot Humid Climate in Villa Rica - Peru: A Study of Thermal Comfort Compared to Confined Masonry, which analyzes the thermal behavior of vernacular architecture compared to confined masonry in hot and humid climates. [22]

Both investigations highlight the importance of recovering and adapting traditional construction techniques depending on specific climatic conditions, promoting more efficient and sustainable alternatives to conventional systems. While the research in Villa Rica evaluates the thermal performance of local materials and designs, the study in Sincos proposes the incorporation of biothermal panels as an innovative solution to improve comfort in cold climates. Together, these works highlight the relevance of passive strategies in Peruvian rural architecture, demonstrating that the combination of ancestral knowledge and modern technology can generate more efficient and resilient living environments.

5 Conclusions

In conclusion, it was determined that the use of natural materials from the area favors the population in terms of low cost and the development of a biothermal panel. They achieve thermal comfort inside homes, the biothermal panel proposal is a sustainable alternative.

The first test that is carried out is with natural materials from the area such as Ichu Fiber, Barley Straw and Plaster, this bio thermal panel has the capacity to retain heat, achieving a space with thermal comfort inside the home, as well as Materials that are local and easy to obtain at low cost are used, such as Plaster. The second test is with Ichu Fiber, Barley Straw and Clay materials, achieving a completely bio thermal panel with materials accessible in the area at low cost, the materials used are thermal, achieving greater thermal comfort, the combination of materials allows for improvement considerably the thermal efficiency properties in the housing prototype which is a performance indicator when applied in real-scale housing.

According to the two tests carried out, the biothermal panels made up of materials from the area are reflected in the fact that the second test has greater thermal capacity, likewise the biothermal panel is accessible and at a low cost for people, giving greater value to the second panel. and being environmentally friendly by reducing pollution.

6 References

1. Castro, A., Davila, C., Wil, L., Cubas, F., Avalos, G., Lopez, C., Menis, L.: CLIMATES OF PERU - National Climate Classification Map. , Red Activa Soluciones Graficas S.A.C. 1(1), 66-68 (2021).
2. National Civil Defense Institute, Frosts affected population centers in the Sincos district, http://sinpad.indeci.gob.pe/sinpad/emergencias/Evaluacion/Reporte/rpt_eme_situacion_emergencia.asp?EmergCode=00077829, last accessed 2023/06/21.
3. Geophysical Institute of Peru.: Climate Atlas of precipitation and air temperature of the Mantaro River Basin. National Environmental Council 1(16), 28-35 (2015).
4. National Institute of Statistics and Informatics.: Peru: Characteristics of private dwellings and homes. 1edn, INEI, Peru (2017).
5. Silva, Y., Mendoca, P., & Maia, C.: Thermal Comfort in Modern Movement – Consequences of the Constructive Definition on the Summer Behavior of a Residential Building in Porto, Portugal., Fourth International Conference on Construction Materials and Materials Engineering, ICBMM 2020, pp. 1-14. Barcelona, Spain (2020).
6. Moncloa, C.: Thermal comfort: An insulating system for high Andean housing made with recycled materials. Architecture Module – CUC 18 (1), 73–90 (2017).
7. Peña, O., & Roman, R.: Design of a thermal insulator based on natural fibers to mitigate the impact of frost in the Cupisa Community. Peruvian University of Applied Sciences, 1-276, (2018).
8. EKOPANELY SERVIS S.R.O, https://www.ekopanely.es/, last accessed 2023/06/12.
9. Ortega, G., Urpeque, D.: Sensorial architecture focused on recreational spatiality. Case of a senior center at the Santiago de Surco Municipality. Cesar Vallejo University, pp. 31-37, (2019).
10. Apaza, J.: Influence of the incorporation of barley straw on the properties of adobe, Coata District, Puno 2022. Cesar Vallejo University, pp. 25-28, (2022).
11. Diaz, C., Diaz, E., Leyva, F.: Production of a lightening element with thermal insulating characteristics based on Gypsum and Rice Husk. National University of Jaén, pp. 20-23, (2021).
12. Mapcarta, https://mapcarta.com/es/25694822, last accessed 2023/05/07.
13. National Civil Defense Institute, http://sinpad.indeci.gob.pe/sinpad/emergencias/Evaluacion/Reporte/rpt_eme_situacion_emergencia.asp?EmergCode=00094180, last accessed 2023/05/07.
14. National Civil Defense Institute, Frosts affected population centers in the Sincos district, http://sinpad.indeci.gob.pe/sinpad/emergencias/Evaluacion/Reporte/rpt_eme_situacion_emergencia.asp?EmergCode=00094180, last accessed 2023/05/07.
15. When to visit, whenvisit.com.pa, https://www.cuandovisitar.com.pa/peru/aramachay-3004673/, last accessed 2023/05/10.
16. Major Academy of the Quechua Language.: Diccionario de la Academia Mayor de la Lengua Quechua. Diccionario de la Academia Mayor de la Lengua Quechua. 2nd edn. AMLQ, Peru (2005).
17. Macek, D., Kosina, M.: Analysis of the structure of heat losses of buildings and benefits of thermal insulation systems. International Civil Engineering and Architecture Conference, CEAC 2021, pp. 01 – 05, Singapore (2020).
18. Dueñas, B., Soto, W., Carrera, E.: Evaluation of the Thermal Efficiency of the Scale Prototype of a Sustainable House that Uses Concrete with PET Fibers (CFP) and the Trombe System. The 4th International Conference on Building Materials and Materials Engineering, ICBMM 2020, pp. 1 - 10, Barcelona, Spain (2020).

 https://doi.org/10.1051/e3sconf/202562003002

19. Gutierrez, D.: Interior surface coverings in rural adobe homes for thermal comfort in areas of extreme temperatures, Puno 2022. Cesar Vallejo University, 1 - 68 (2022).

20. Montoya , V. Ancestral Earthen Construction Techniques Updated to the Needs of the People in the Central Andes of Peru, an Experience of Research and Training of Architecture Students Based on Community Service. En World Congress of Architects UIA 2023 . Cham: Springer International Publishing, 2023. p. 307-315.

21. Salazar A., et al. Lightened Blocks of Plant Fibers and Clay Applied in Construction Systems for Housing in High Andean Areas. En E3S Web of Conferences. EDP Sciences, 2024. p. 04002.

22. Castro , N.; Montoya, V. Revaluation of Vernacular Architecture in a Hot Humid Climate in Villa Rica-Peru: A Study of Thermal Comfort Compared to Confined Masonry. 2024. Civil Engineering and Architecture, Vol. 12, No. 6, pp. 3888 - 3903, 2024. DOI: 10.13189/cea.2024.120611.

Effect of Repair and Reuse on small WEEE Generation Rates in Australia

Gimhan Jayasiri[1*], *Sunil* Herat[1], and *Prasad* Kaparaju[1]

[1]School of Engineering and Built Environment, Griffith University, Australia

Abstract. The escalating generation of small Waste Electrical and Electronic Equipment (WEEE) poses a significant global challenge, constituting 30% of total WEEE in 2019. Projections suggest a 90% increase in Australia over the next 25 years. This study delves into the impact of repair and reuse on small WEEE, examining the interplay between product lifespan, products on the market (PoM), and waste generation rates (WGR). The findings from a Waste Generation Estimation model underscore that repair and reuse efforts, in isolation, have a limited short-term impact on WGR reduction. A substantial decrease in WGR can only be achieved by reducing the PoM, driven by Repair and Reuse. This underscores the necessity of a comprehensive approach, which involves incentivizing consumers and placing greater responsibility on producers to manage the entire lifecycle of small EEE. By implementing strategies such as eco-modulation and takeback schemes, alongside promoting durability and repair, the Circular Economy can be better integrated into managing this growing waste stream.

1 Background

Electrical and Electronic Equipment (EEE) refers to devices that contain circuitry and components powered by batteries or an external power source, encompassing various items used in households and businesses [1]. Increased consumption rates and reduced lifespans significantly rise in Waste Electrical and Electronic Equipment (WEEE), the fastest-growing solid waste stream globally. [2]. In 2022, 62 billion kilograms of WEEE was generated worldwide, an average of 7.8 kg per capita annually. Oceania ranked as the second-highest region regarding WEEE generation, with an average of 16.1 kg per capita in the same year[3]. Despite the need for proper WEEE management, most EEE is still disposed of or recycled uncontrolled [3-5]. Hence, a proactive approach is essential for managing this growing waste stream, incorporating principles of a sustainable Circular Economy. Within the waste management hierarchy, repair and reuse play a key role in reducing waste by prolonging the lifespan of products. The decision to repair or reuse EEE is influenced by several factors, including energy efficiency [4], the type of repair needed [5], the quality of repair [6], consumer attitudes towards reusing repaired EEE [7, 8], and the ease of product disassembly [9]. Numerous studies have explored the environmental benefits of repairing and reusing EEE,

* Corresponding author: gimhan.jayasiri@griffithuni.edu.au

https://doi.org/10.1051/e3sconf/202562004001

focusing on extending product lifespans through Life Cycle Assessment (LCA) [4, 10-12]. These studies have examined large appliances and IT products, such as washing machines and laptops. Given the scarcity of studies comparing the impact of repair and reuse across different categories of EEE [13, 14], a comprehensive numerical model is necessary to understand their effect on WGR better. This study addresses this gap by analysing the scenario specifically for small EEE, which accounted for over 30% of global WEEE generation in 2019 [15] and 2022 [3]. Australia was chosen as the geographical focus for this study, as it produced 577 kt of WEEE in 2021, with small WEEE accounting for 172 kt (approximately 30% of the total). Moreover, Australia's WEEE generation rate of 22.4 kg per capita is comparable to that of many European nations, which are among the world's largest generators of WEEE.[16].

In examining Australia's WEEE management schemes, the National Television and Computer Recycling Scheme (NTCRS) is the primary co-regulatory initiative under the Product Stewardship Act. Its primary goals are to reduce electronic waste in landfills, enhance material recovery, and provide recycling access to households and businesses. Importers and manufacturers are held responsible for their products' entire lifecycle and must join Producer Responsibility Organizations (PROs) under NTCRS, which organize WEEE collection and recycling on behalf of their members [17]. However, there is currently no dedicated scheme to manage small WEEE in Australia, although discussions are underway to implement such a program [18].

In Australia, small EEE are frequently disposed to landfill or sent to recycling plants due to a lack of a strong repair market [19] and the use of low-quality materials that reduce product lifespans [20]. Despite 85% of Australians valuing durability and repairability in products, many small appliances like toasters and blenders are replaced after their first malfunction [21]. Manufacturers often hinder repairs by not providing necessary equipment or spare parts, and repair costs can rival replacement prices, making repairs economically unattractive it [19]. While most states allow unlicensed individuals to repair low-voltage household EEE, some of the product designs make self-repair challenging [19]. Community interest in repair and reuse is growing, supported by repair cafés and tool libraries offering affordable solutions [21]. However, the rise of IoT-connected devices and manufacturers' planned obsolescence strategies may further decrease repair opportunities. Consumers are increasingly concerned about the environmental impact of short-lifespan products and are advocating for more sustainable practices [19].

2 Methodology

2.1 What are small EEE?

Standardising management practices involve classifying EEE based on function, material composition, average weight, and lifespan. The United Nations University has developed a system called UNU-keys for EEE, which are also associated with harmonised statistical (HS) codes [15]. Currently, EEE is divided into six categories, with small EEE being the most significant contributor to global WEEE generation. This category typically includes small household appliances and devices. In the Australian context, 17 UNU-Keys have been identified as small EEE, as outlined in Table 1 below.

Table 1. Small EEE in Australia [22].

UNU Key	Typical Appliances	Average Lifespan (yrs)	Average Weights (kg)
0114	Domestic microwave ovens	15.9	21.5
0201	Electric blankets, roof fans, personal scales, household sewing machines, clocks, and watches	7.5	1.06
0202	Domestic food processors, mixers, and toasters	9.8	3.27
0203	Electric water heaters, coffee makers, and tea makers	7.2	1.89
0204	Vacuum cleaners	9.9	5.67
0205	Hair dryers, shavers, and hand dryers	7.6	0.55
0401	Microphones, headphones, and speaker sets	9.1	0.39
0402	Portable radios and broadcast receivers without external power	9.0	0.27
0403	Amplifiers, musical instruments, and office machines	8.9	3.82
0404	Video recorders, DVD players, and projectors	7.9	3.51
0405	Single and multi-speaker loudspeakers	12.0	2.33
0406	Digital cameras and video recorders	6.4	0.45
0601	pressure washers, lawnmowers, sanders, hedge trimmers, leaf blowers, and electric chainsaws	13.3	2.52
0701	Electric trains, racing sets, and drones	4.1	0.45
0702	Gaming consoles	4.6	0.48
0801	Blood pressure monitors, hearing aids, glucose meters, fitness trackers, and digital pulse oximeters.	11.9	0.18
0901	Smoke detectors, alarms, and measuring instruments	5.3	0.24

2.2 Modelling approach

Estimating future WEEE generation rates is crucial for determining appropriate management strategies. Despite Australia being one of the largest WEEE generators, only two academic studies [23, 24] and three consultant-developed models[25-27] have been developed for the country. Globally, Material Flow Analysis (MFA) is the most used method for modelling WEEE generation, with both dynamic and static MFA techniques available. Dynamic MFA, which utilises long-term data, provides deeper insights for policymakers, while static MFA estimates WEEE generation for a single year. There are two main approaches for WEEE generation modelling: stock-based and sales-stock-lifespan-based. Given the availability of sales data and its importance in predicting future WEEE quantities, this study used the dynamic MFA method with a sales-stock-lifespan model to estimate small WEEE generation rates in Australia. The model was developed using the Python programming language.

2.2.1 Variables and parameters

Modelling WEEE generation rates depends primarily on product lifespan and the number of items in the market. For small EEE, products are often discarded before reaching their lifespan due to varying consumer expectations. Most studies use fixed average lifespan inaccuracies [28, 29], which needs to capture the dynamic nature of product usage. To address this, the Weibull distribution function is commonly used in MFA models to reflect gradual failure rates. Despite limited consumer data in Australia, this study utilized UNU data to create lifespan profiles for small EEE, applying equation 1 to estimate their lifespan.

$$L(p)(t,n) = \frac{\alpha}{\beta^{\alpha}}(n-t)^{\alpha-1}e^{-(\frac{n-t}{\beta})^{\alpha}} \quad (1)$$

where,
α, β – shape and scale parameters
n – evolution year
t – historical year
$L(p)(t,n)$ – Lifespan of the small EEE

The mass of small EEE placed on the market (PoM_t) is a crucial factor that influences the WGR. This study used the apparent consumption method indicated by equation 2, based on UNU guidelines, to calculate (PoM_t) considering Imports (I_t) and Exports (E_t), as domestic production of small EEE in Australia is negligible.

$$PoM_t = I_t - E_t \quad (2)$$

Import and export data was extracted from the UN Comtrade database for 1992-2022, and linear interpolation was used to correct the discrepancies. Packaging content was estimated at 38.10%, following European guidelines [30]. Using this approach, the PoM of small EEE in Australia from 1992 to 2022 was calculated.

2.2.2 Forecasting methods

According to UNU guidelines, sales data (PoM_t) is more sensitive than lifespan values, and selecting an accurate forecasting method is crucial. Literature suggests logistic regression for stock-based models[31, 32] and linear regression for sales-based studies [33] , though these have not accounted for population growth's effect on PoM. The study used existing per capita PoM data and forecasted future ratios based on Australia's population projections to address this. Exponential smoothing techniques were applied since no linear relationship was found between PoM and the 17 small EEE categories. After time series decomposition, the Double Exponential Smoothing method [24] was used to forecast PoM for a further 28 years. Then using the lifespan profiles along with current and projected PoM data for small EEE, the generation rates of small WEEE were calculated based on Equation 3.

$$WGR\ small\ WEE_n = \sum_{t=1992}^{n} PoM_t \times L(p)(t,n) \quad (3)$$

2.2.3 Applying the effect of repair and reuse

After calculating the WGR for small EEE, the impact of repair and reuse on these rates was modelled by adjusting five variables: lifespan, the percentage and year of small EEE repaired or reused, the percentage reduction in PoM, and the year this reduction is applied.

3 Results and Discussion

3.1 Reference model

The WGR for small EEE in Australia is projected to increase by 90% from 197 kt/yr in 2025 to 368 kt/yr by 2050, driven by rising consumption, as shown in Figure 1. However, specific categories like Clocks, Watches, Irons, Fans (UNU Key 0201), Portable AV (0402), and Video Players (0404) are expected to see a decline in WGR due to reduced production and the replacement of outdated products with more advanced devices. In contrast, Toys (UNU Key 0701) remain the most significant contributor to WGR, followed by food preparation EEE (0202) and household tools (0601).

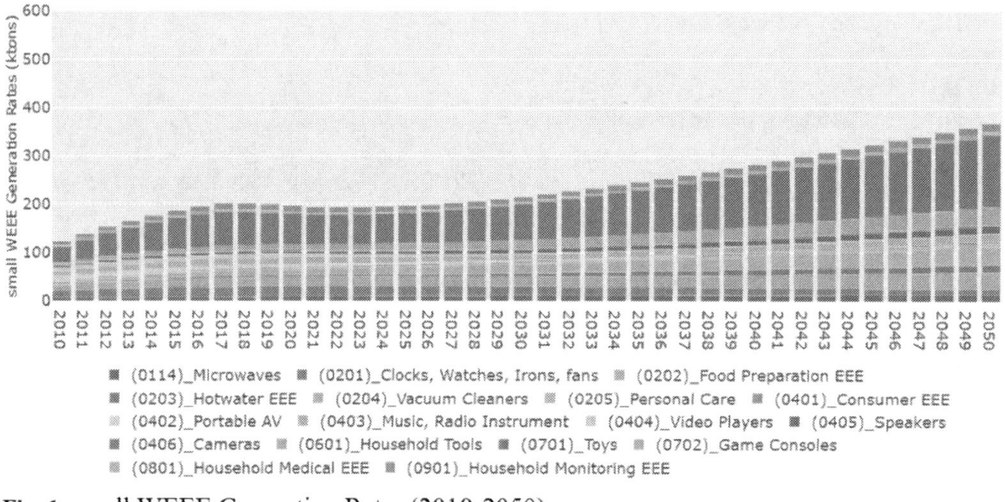

Fig. 1. small WEEE Generation Rates (2010-2050)

3.2 Effect of Repair and Reuse

The study quantitatively modelled the impact of repair and reuse on small WEEE generation rates in Australia. The findings suggest that extending the lifespan of small EEE has a more substantial short-term effect on reducing WGR than its long-term impact. The lifespan was varied during the initial part of the study without changing the PoM small EEE. For example, a 20% lifespan increase combined with 50% of products being repaired and reused reduces

WGR by 3.9% in 2035 and 2.25% in 2050. The most significant reduction, 16.2% in 2035 and 10.6% in 2050 occurs when all products are repaired and reused, with a 50% lifespan increase. Conversely, a shorter lifespan leads to a more significant rise in WGR, almost double the decrease observed with lifespan extensions. During the study's second stage, the lifespan and PoM of small EEE varied. It was observed that the reduction in small WEEE generation rates is much more significant when the PoM is reduced in parallel with an increase in lifespan. This indicates that small EEE is highly responsive to these combined efforts. This trend persists across different scenarios, even when the lifespan is increased substantially.

This study also modelled scenarios where the lifespan is shortened due to the early discontinuation of small EEE, where there are few repaired and reused small EEE. The pattern was similar as the WGR increased significantly with the increase in PoM. Overall, the findings suggest that, regardless of whether the lifespan is increased or shortened, the resulting reduction in WGR is affected mainly by the variations in PoM.

It is important to note that scenarios involving 100% repair and reuse were not considered, as such an event is improbable. This further highlight that a significant reduction in the WGR of small EEE can be achieved when high levels of product reuse are coupled with a decrease in PoM. However, reaching these higher rates of PoM reduction may prove challenging due to current economic trends and market conditions. Some of the critical outputs of the model are highlighted in Table 2 below.

Table 2. Effect of Repair and Reuse on small WEEE Generation Rates

Life Span Increase (+) or Decrease (-)	Affected small EEE (%)	The initial year of the change	PoM Increase (+) or Decrease (-)	Initial year of the change	Increase/ Decrease in small WEE Generation rates (%)			
					2035	2040	2045	2050
20%	25%	2030	-25%	2030	-14%	-21%	-24%	-25%
20%	50%	2030	-50%	2030	-28%	-42%	-47%	-50%
-20%	25%	2030	25%	2030	16%	23%	25%	26%
-20%	50%	2030	50%	2030	34%	47%	51%	52%
20%	25%	2030	-25%	2030	-14%	-21%	-24%	-25%
50%	25%	2030	-25%	2030	-16%	-23%	-26%	-26%
-20%	25%	2030	25%	2030	16%	23%	25%	26%
-50%	25%	2030	25%	2030	23%	28%	28%	28%

4 Conclusions and Recommendations

In conclusion, this research highlights the dynamics of small WEEE generation rates in Australia and provides essential insights into future trends and interventions of the Circular Economy. The variation in WGR under different scenarios emphasises the complexity of predicting the outcomes of repair and reuse. While repair and reuse have a more noticeable short-term impact, substantial reductions in WGR require increased product reuse and reduced small EEE placed on the market (PoM). This was also confirmed in other studies that PoM is more critical in reducing WEEE generation rates compared to product lifespan[34]. Achieving significant PoM reductions, however, may prove difficult due to current economic conditions.

The results show that without PoM reduction, the effect of repair and reuse on WGR is limited. Manufacturers should, therefore, extend their Producer Responsibility programs to cover the entire product lifecycle, from production to disposal. Remanufacturing which is another form of reuse is recommended which reduces the production of new products [35]. Given the minimal domestic manufacturing of small EEE [26], regulating imports through eco-modulation and a product index [36] could help extend product lifespan by promoting durability and easier repair.

Incentives such as reward points, extended warranties, and takeback schemes could encourage consumers to keep and repair small EEE rather than discard them. Producers should also conduct economic evaluations across the product lifecycle to maximise material circularity and determine optimal production levels. Although consumer interest in small EEE repair and reuse is low, manufacturers are primarily responsible for promoting these practices. Addressing these issues will become increasingly critical as small WEEE generation rates grow and E-waste landfill bans loom.

References

1. Withanage, S.V. and K. Habib, *Life Cycle Assessment and Material Flow* Analysis*: Two Under-Utilized Tools for Informing E-Waste Management.* Sustainability, 2021. **13**(14): p. 7939.
2. N. Perkins, D., et al., *E-Waste: A Global Hazard.* Annals of Global Health, 2014. **80**(4): p. 286.
3. Cornelis P. Baldé, R.K., Tales Yamamoto, Rosie McDonald, Elena D'Angelo, Shahana Althaf, Garam Bel, Otmar Deubzer, Elena Fernandez-Cubillo, Vanessa Forti, Vanessa Gray, Sunil Herat, Shunichi Honda, Giulia Iattoni, Deepali S. Khetriwal, Vittoria Luda di Cortemiglia, Yuliya Lobuntsova, Innocent Nnorom, Noémie Pralat, Michelle Wagner *The Global E-Waste Monitor 2024.* 2024, United Nations Institute for Training and Research.
4. Pérez-Belis, V., et al., *Environmental performance of alternative end-of-life scenarios for electrical and electronic equipment: A case study for vacuum cleaners.* Journal of Cleaner Production, 2017. **159**: p. 158-170.
5. Bovea, M.D., et al., *Potential reuse of small household waste electrical and electronic equipment: Methodology and case study.* Waste Management, 2016. **53**: p. 204-217.
6. Dindarian, A., A.A.P. Gibson, and J. Quariguasi-Frota-Neto, *Electronic product returns and potential reuse opportunities: A microwave case study in the United Kingdom.* Journal of Cleaner Production, 2012. **32**: p. 22-31.
7. Bovea, M.D., et al., *A survey on consumers' attitude towards storing and end of life strategies of small information and communication technology devices in Spain.* Waste Management, 2018. **71**: p. 589-602.
8. Pérez-Belis, V., M.D. Bovea, and A. Simó, *Consumer behaviour and environmental education in the field of waste electrical and electronic toys: A Spanish case study.* Waste Management, 2015. **36**: p. 277-288.

9. Li, J., M. Barwood, and S. Rahimifard, *A multi-criteria assessment of robotic disassembly to support recycling and recovery.* Resources, Conservation and Recycling, 2019. **140**: p. 158-165.
10. Schischke, K., et al., *Life cycle energy analysis of PCs–Environmental consequences of lifetime extension through reuse.* power, 2003. **4**(5.248): p. 5.032.
11. Lu, B., et al., *Comparison on End-of-Life strategies of WEEE in China based on LCA.* Frontiers of Environmental Science & Engineering, 2017. **11**: p. 1-12.
12. Zink, T., et al., *Comparative life cycle assessment of smartphone reuse: repurposing vs. refurbishment.* The International Journal of Life Cycle Assessment, 2014. **19**: p. 1099-1109.
13. Pini, M., et al., *Preparation for reuse activity of waste electrical and electronic equipment: Environmental performance, cost externality and job creation.* Journal of Cleaner Production, 2019. **222**: p. 77-89.
14. Bovea, M.D., V. Ibáñez-Forés, and V. Pérez-Belis, *Repair vs. replacement: Selection of the best end-of-life scenario for small household electric and electronic equipment based on life cycle assessment.* Journal of Environmental Management, 2020. **254**: p. 109679.
15. Vanessa Forti, et al., *The Global E-waste Monitor 2020.* 2020, United Nations Institute for Training and Research.
16. OECD. *Waste from electrical and electronic equipment.* 2021 31.10.2024]; Available from: https://data-explorer.oecd.org/vis?tenant=archive&df[ds]=DisseminateArchiveDMZ&df[id]=DF_EWASTE&df[ag]=OECD&dq=AUS.EE6%2BEE_SE..&pd=2000%2C2021&to[TIME_PERIOD]=false&vw=tb.
17. Australian Government and Department of the Environment, *The National Television and Computer Recycling Scheme —Operational Review.* 2014.
18. Department of Agriculture and Water and Environment, *Stewardship for Consumer and Other Electrical and Electronic Products (Discussion Paper).* 2021.
19. Productivity Commission, *Right to Repair - Productivity Commission Inquiry Report.* 2021, Australian Government.
20. Banaszkiewicz, K., et al., *Household E-Waste Management: A Case Study of Wroclaw, Poland.* Sustainability (Switzerland), 2022. **14**(18).
21. Office of Circular Economy and Department of Environment and Science, *Queensland E-Products Action Plan 2023 - 2033 (Draft)*, E.a. Science, Editor. 2023.
22. Jayasiri, G., S. Herat, and P. Kaparaju, *Repair and Reuse or Recycle: What Is Best for Small WEEE in Australia?* Sustainability, 2024. **16**(7): p. 3035.
23. Golev, A., et al., *Where next on e-waste in Australia?* Waste Management, 2016. **58**: p. 348-358.
24. Islam, M.T. and N. Huda, *E-waste in Australia: Generation estimation and untapped material recovery and revenue potential.* Journal of Cleaner Production, 2019. **237**: p. 117787.
25. Randell, P., *Victorian e-waste material flow analysis 2021.* 2022, Randell Environmental Consulting,
Blue Environment.
26. Bontinck PA, et al., *E-product stewardship in Australia: Evidence report. A report prepared for the Commonwealth Department of Agriculture, Water and the Environment by Iceni Group and Lifecycles, Sydney, Australia.* 2021.

27. Busby G and Bremner A.M, *Western Australia E-Waste Material Flow Analysis*. 2021, Encycle Consulting.
28. Robinson, B.H., *E-waste: an assessment of global production and environmental impacts.* Science of the total environment, 2009. **408**(2): p. 183-191.
29. Ikhlayel, M., *Differences of methods to estimate generation of waste electrical and electronic equipment for developing countries: Jordan as a case study.* Resources, Conservation and Recycling, 2016. **108**: p. 134-139.
30. European Committee of Domestic Equipment Manufacturers (CECED), *Material Flows of the Home Appliance Industry*. 2018.
31. Tran, H.P., et al., *Material flow analysis for management of waste TVs from households in urban areas of Vietnam.* Resources, Conservation and Recycling, 2018. **139**: p. 78-89.
32. Habuer, J. Nakatani, and Y. Moriguchi, *Time-series product and substance flow analyses of end-of-life electrical and electronic equipment in China.* Waste Management, 2014. **34**(2): p. 489-497.
33. Parajuly, K., K. Habib, and G. Liu, *Waste electrical and electronic equipment (WEEE) in Denmark: Flows, quantities and management.* Resources, Conservation and Recycling, 2017. **123**: p. 85-92.
34. Forti V., B.C.P., Kuehr R, *E-Waste Statistics: Guidelines on classification, reporting and indicators*. 2018, United Nations University, IAS - SCYCLE, Bonn, Germany.
35. King, A.M., et al., *Reducing waste: repair, recondition, remanufacture or recycle?* Sustainable Development, 2006. **14**(4): p. 257-267.
36. Jayasiri, G., S. Herat, and P. Kaparaju, *Management of Small WEEE: Future Directions for Australia.* Sustainability, 2023. **15**(18): p. 13543.

https://doi.org/10.1051/e3sconf/202562004002

The effects of conflict pollution on environmental sustainability: A modern adaptation of the just war theory to foreign policy as a method of pollution prevention

Daisy Spiva[1]
[1]University of North Carolina Wilmington, United States

Abstract. As it stands, the environment is confronted by an unprecedented extent of human activity from ongoing technological innovation, population growth, and resource consumption. This current state of development isn't sustainable without interventive measures to mitigate risks of ecosystem destabilization and eventual collapse. Though human impact on the environment isn't necessarily negative, these perturbations in ecological conditions are of substantial concern due to an accelerated rate of modification and degradation that compromise geological processes. Historically, Earth and its environment underwent several periods of change, such as glacial-interglacial cycles from astronomical forces; however, these fluctuations are natural constituents of planetary development and the phase of slow surface evolution. In contrast, the inception of conflict is an additional and artificial disturbance with the propensity to cause catastrophic effects, despite strategies to counteract the harmful byproducts of societal operations. In terms of air, water, and soil resources, outbreaks of conflict are a potential source of rampant pollution and their environmental costs are referenced within international law with limited protection. Regardless of outlined provisions that prohibit the direct destruction of the environment as a weapon of war, contemporary advancements in military organization and their technological capabilities put the environment at greater risk than ever before in human history. Periods of active combat also undermine environmental mitigation initiatives by allocating the necessary funding, research, and workforce elsewhere due to the inherently demanding nature of conflict. Thus, I argue a modern adaptation to the Just War Theory and its application in foreign policy can further inform efforts of environmental protection and pollution prevention, ensuring global sustainability as well as forestalling the onset of irreversible damage.

1 Introduction

The notion of environmental sustainability adopts different forms and approaches, depending on the type and degree of hazards, which requires intervention methods to be in a constant flux of customization to ongoing trends. The environment is far from static but exceptionally dynamic from continuous changes in natural and human processes that exert influence over landscapes along with their inhabitants. More specifically, the current geological epoch of the Anthropocene is characterized by the extensive impact of human activity on the composition of and interaction between elements within the ecosphere. As human civilizations advanced in terms of infrastructure and production, the range of environmental consequences from various human activities expanded to an unprecedented magnitude. Modernization transforms the global environment, especially through pollution, by human input disrupting initial ecological conditions as stressors that compromise ecosystems' structure, function, and quality. The framework of human-environment interaction involves diverse components, from technological inventions to dispersal patterns, that together dictate the outcomes of our relationship with the planet. This feedback loop positions society to not only depend on and adapt to the environment but also modify it, which then prompts ecosystems to undergo alterations that negatively impact their ability to regulate as well as thrive. Similarly, the expansion of scale and acceleration of pace from modernizing systems is of detriment to nature's capacity to perform the ecological resistance and resilience necessary for sustainable progress. Human culture and innovation will only continue to exhibit the aptitude to reorient the natural course of planetary development and health, with these perturbations having crucial ramifications for the Earth's overall habitability. The resulting transformations from our ecological footprint, if not mediated through a holistic approach to intervention, eventually will cause vast environmental damage and some of which may be irreversible.

Furthermore, the concern of warfare potential in the contemporary era is a sort of human activity that cannot only undermine environmental initiatives and programs but simultaneously function as a new source of pollution to account for. Air, water, and soil resources, which are imperative to natural processes, historically have been exposed to a wide array of pollutants from military facilities, modes of transportations, and equipment. At that, international agreements only offer limited protection to the environment under the extenuating circumstance of conflict, more so regulating its direct weaponization and courses of action that result in "widespread, long-term, and severe damage." [1] However, aside from extreme hostile use of environmental modification as an attack or reprisal, the environment still suffers consequences from outbreaks of conflict that technically fall within the scope of guidelines mentioned in international law. The conflict pollution that isn't deemed to fit the description of what is explicitly prohibited remains concerning in its own right and can be furthered by variables of warfare, such as frequency, duration, and quantity of parties. To amend the disparity between policy and practice to better uphold sustainability, I argue modern adaptations of the Just War Theory are valuable to remedying the limits to environmental protection within international law. Given the state of modernization in human society, present and future conflicts must recognize the environmental costs that were once justifiable under conditions that predate the latest technological advancements, population growth, and resource consumption aren't equally practical to the world of today. In other words, if the environment is in a more compromised position now than during past wars, then the standard of justification for conflict should ultimately be higher and the permissibility of its environmental costs ought to reflect that. The criteria of the Just War Theory already contributed insight to foreign policy, especially in reference to the principle of proportionality, yet its application can nonetheless benefit from amendments that better reflect the implications of modernity on the human-environment interaction.

2 Planetary Development, Health, and the Anthropocene

In order to address planetary health in the contemporary era, it's functional to frame those developments in relation to the origins of Earth and our solar system as a whole. Once the solar nebula, a molecular cloud of dust and gas, collapsed under its own gravity, the immense pressure on the interstellar mass stimulated nuclear fusion that prompted the formation of the Sun. With the densest mass towards the center, otherwise within the Sun's core, the remaining material "was flattened, due to rotation, forming a disc" that over time coalesced into various celestial bodies including Earth. [2] Through the process of accretion, particles of the solar nebula gradually integrate from gravitational attraction, with the denser materials settling to the core and finer materials rising to the surface during the stage of differentiation. Then the cratering stage comprises frequent collisions with celestial bodies due to the sheer amount of space debris still present, which bombard and fracture the newly forming crust. In turn, the flooding stage ensues when magma in the mantle flows out of fissures at points of impact from cratering and fills residual basins. Additionally, said crater basins are later flooded with water after water vapor released into the atmosphere from the same fissures "condensed and started to fall as rain, forming the oceans and other liquid waterbodies on the surface." [2] Due to Earth's positioning within a habitable zone of space, in which the temperature range permits water to exist in a liquid state, life developed with water acting as a medium for its essential chemical reactions. This particular trajectory of Earth's evolution was largely determined by pre-existing variables and constraints, such as proximity to the Sun, course of orbit, and initial exposure to certain elements. In time, subsequent geological processes and the genesis of living organisms factored into the stabilization of the environment during the final stage of slow surface evolution. Aside from the presence of liquid water, other features of Earth's habitability include an atmospheric magnetic field for the deflection of solar radiation, plate tectonics for climate regulation, and a balance of chemical elements for the building blocks of life.

As previously mentioned, the evolutionary process towards extant ecosystems and organisms required a certain series of preceding ecological conditions, which modern human activity endangers due to factors like industrialization, resource depletion, overpopulation, and urbanization. There is a substantial risk to the environment since anthropogenic disturbances compromise both abiotic and biotic elements within ecosystems at an increased rate and scale. The human-environment interaction entails a feedback-loop based on adaptation, dependency, and modification that acts as an ongoing source of perturbations that, if not regulated, can surpass environmental thresholds. Historically, this relationship led to exploitation and a decline in ecosystem health since "humans control the ecosystem attributes that are valued, they are often the source of stressors that push systems toward and over threshold." [3] Even minor disturbances in ecological conditions have the potential to transform into a consequential rapid change once the threshold is exceeded, which undermines the ecosystem's practice of resistance and resilience as a means to sustain itself. Ecological resistance refers to the ability to withstand disturbances and preserve the current state of existence, while ecological resilience is the ability to recover from said disturbances without any loss of structural integrity or shifts in function. These two concepts operate concurrently to offer necessary bandwidth to the environment, promoting receptivity to fluctuations and deterring radical destabilization capable of ecosystem collapse. In order to prevent decentring the environment from alignment with natural processes that specifically evolved relative to prior conditions, there must be orders of protection that assess threshold capacity in proportion with the unprecedented influence of modern human activity. Comparatively, there also needs to be emphasis on the preservation of the environment's ability to engage in ecological resistance and resilience to adapt on its own volition outside human involvement.

3 Role of Conflict and its Environmental Costs

In particular, the inception of conflict is increasingly concerning since military modernization has advanced forms of organization, training, equipment, and doctrine that escalate the potentiality of greater environmental costs. Since the military is another facet of society, the progression of human civilization over time led to a multitude of developments, especially in terms of mechanization and robotics, that alter warfare potential and the materialization of conflict in the contemporary landscape. For example, emerging technologies precipitated a rise in the mass production of weaponry and global accessibility to nuclear devices, all of which are more powerful than their precedents in human history. Advanced weapon systems also incorporate automation and artificial intelligence to increase efficiency by streamlining operations and optimizing response times. Though the environment suffered from the destruction and pollution of conflict in ancient times, the state of modern technology furthers such consequences to a degree that causes war to be more unsustainable for our environment than ever before. The progressive upgradation of military research and services led to the establishment of an unprecedented network of technologies specifically designed to enhance combat capabilities and, therefore, disturb pre-existing ecological conditions. In addition to extensive exposure to anthropogenic disturbances related to the nonviolent social and economic endeavors compulsory for societal functioning, the onset of simultaneous wars or another world war can surely exceed environmental thresholds and risk ecosystem collapse. Even if war is at times a necessary countermeasure to grievances or brutality, its environmental costs may still not be practical if the resulting harm to the environment jeopardizes Earth's habitability. Meaning, if the environment isn't able to endure additional disturbance from decisions to engage in conflict, regardless of the reasons underpinning that determination, then the possible danger of causing the world to become uninhabitable for the entire population subverts the humanitarian cause to mediate detriment between select parties.

Since humans are in the process of environmental assessment and management to mitigate the effects of existent pollution, redirecting the allocation of resources to mobilization not only functions as a disinvestment to the aforementioned initiatives but also furtherance of the target concern. If there is already expressed difficulty in the regulation of human activity and its ramifications on the environment, then we might not be able to afford the counterproductive nature of deeming war permissible during a time of ample environmental vulnerability. Nonetheless, the recent developments in international affairs within the last several decades attest to an era that is capable of having conflict along with its environmental costs, despite the compounding risk of uninhabitability and irreversible damage. According to the global peace index (GPI), the world is currently experiencing the highest rates of conflict since World War II at a "total of 56 incidents around the globe" and "92 countries engaged in conflicts outside their own borders." [4] These statistics are unprecedented in the history of GPI reports, with there being the most countries ever recorded to participate in cross-border conflict as well as the most countries (108) to undergo further militarisation. [4] Among some of the most notable wars are the Russo-Ukrainian conflict and Israeli-Palestinian conflict, which have heightened tensions on the international scale regarding what political and economic strategies are suitable to address events of injustice. Likewise, there are several civil wars occurring in countries like Sudan, Myanmar, and Syria, furthering the sense of global instability as military spending and operations are on the rise. Evidently, these trends of insurgency and invasion incite alarming episodes of environmental degradation from coinciding pollution, which can have persisting effects in the future even after conflict is resolved. Though the pressures to engage in conflict are not always preventable, their means of resolution is a choice which must entail informed decision-making that appropriately gauges whether the benefit justifies the cost, including the possibility of Earth's uninhabitability.

4 Just War Theory in the Modern Era

Outbreaks of conflict have plagued human history for millenniums, with the earliest recorded war occurring within the Neolithic era in Mesopotamia between Sumer and Elam around 2700 BCE. However, this still doesn't account for wars that predate recorded history and the archaeological evidence of remnants from potential "mass killings" in the Mesolithic era, which to some suggest an interminable relationship between human beings and waging wars. [5] Regardless, over time the proclivity of conflict among human civilizations prompted certain individuals to conceive of philosophical rationales and military doctrines that regulate the destructive, violent nature of war. More specifically, the Just War Theory was explicitly devised to provide ethical instruction to the relevance and application of war through outlining in a set of criteria the necessary constituents for its justification. As a means to moderate its frequency and prevent war from becoming excessive, the Just War Theory frames conflict in terms of moral boundaries that legitimize some cases while limiting others. Yet the periodicity of conflict is subject to the extenuating circumstances of social organization, especially when modernization has drastically altered the state and scale of war from times of prehistoric warfare. Though the Just War Theory is a framework that aims to determine the defining characteristics of conflict justification, it's on the presumption that war is at times necessary; meanwhile, recent developments shift to question whether war is a sustainable option irrespective of said necessity. Given the current circumstance of global environmental vulnerability due to ongoing processes like industrialization and urbanization, the consequent degradation from conflict may be inconsistent with the overall objective of the Just War Theory in light of the growing destructive power of technological advancements.

There are two groupings of criteria, jus ad bellum and jus in bello, with the former concerning the right to engage in war and the latter concerning appropriate conduct during war. The principles of this theory originated from the likes of Plato and Cicero while eventually being further developed by later philosophers, such as St. Augustine and Thomas Aquinas, inevitably each time evolving in reference to their respective eras and circumstances. [6] Despite a diverse range of investigations reflecting shifts in human condition and culture across numerous nations, there remain notions like comparative justice, probability of success, and war as a last resort that are relevant in the majority of cases. In regard to the environmental costs of modern warfare, I intend to elaborate on the proportionality of ends in jus ad bellum and the proportionality of means in jus in bello. The principle of proportionality entails that the outcome of conflict should foster a better state of affairs and any harm caused from the use of force should be consistent with the desired goal, all the while actively minimizing the extent of mass destruction and loss of life during active combat. Yet the execution of proportionality is contingent upon the relations between political orientations, military tactics, and technological resources, which all influence the configuration and yield of conflict. Withal, the ultimate goal, according to the Just War Theory, is the re-establishment of peace and that to the extent where the peace following war is preferable to any semblance of peace prior to it. However, the query of whether conflict is justifiable depends on if modern wars are capable of achieving the intended results of peace and proportionality. For example, a nation declaring war in response to acts of brutality can succeed in deterring future attacks and that restoration of justice can still ultimately be short-lived; if said conflict also triggers environmental harm to the extent of jeopardizing planetary habitability, then those saved along with the rest of the population are at an even greater risk due to the simultaneous destruction of any viable expanse to promote peace on. Thus, the risk of war may appear reasonable if limited to the lens of the human input it's reactive to but may not in the scope of war's environmental costs threatening the livelihood of the planet and human species alike.

Though the relevance of Just War Theory is questioned and deemed outdated by some, the concept of proportionality remains indispensable to international humanitarian law (IHL). Apart from the other criteria of legitimate authority, right intention, just cause, and discrimination, proportionality differs since it intrinsically assesses acts of war in conjunction with their after-effects. [6] Meaning, the resulting phenomena of conflict, or respectively its environmental costs for the purpose of this argument, are integral to determining permissibility. If there is a requirement of balance between means and ends, then the justification of modern warfare is debatable given the possibility of Earth's uninhabitability as a feasible outcome. The evolution of military forces involves an exponential increase of destructive power, weapon accessibility, supportive equipment, and resource consumption, all of which further the severity of environmental costs. Centuries ago, warfare was characterized by swords, spears, and arrows with soldiers traveling either by foot or horse; however, militaries now utilize assault weapons with large-capacity magazines, projectile missiles, and nuclear explosives with armored vehicles and a variety of aircrafts at their disposal. The historical span of scientific and technological developments led to the incorporation of factors once non-existent, from chemical warfare agents to lethal autonomous weapons, causing warfare potential to take on a deadlier form that can more easily incite ecosystem collapse. Additionally, the environmental impact from the introduction of these advancements in conflict not only prompted greater lengths of destruction but new sources of pollution. Whether it be the extensive use of fossil fuels to power weapon systems or the ensuing generation of greenhouse gas emissions from said practice, military modernization has fundamentally changed how war is waged and the breadth of its environmental costs.

Though the threat of the planet becoming uninhabitable was highly unlikely during times of prehistoric warfare, products of innovation have altered the previous propensity of harm from acts of war and, "given the deadliness of current weaponry and the deficiency of coherent moral rationalisation for engaging in any war, it is time for a severe rethink." [7] Moreover, the scope of degradation the environment has been exposed to since the first Industrial Revolution to the Technological Revolution of today problematizes the ability of war to produce improvements to the state of affairs, since the onset of conflict further compromises environmental wellness and the status of habitability. In other words, there is no possibility of a truly just war if it risks destabilization and irreversible damage in an already threatened environment. There is no sense of proportionality or peace if a war is waged, regardless of its reasoning, and Earth becomes uninhabitable putting the entire global population in peril. The possibility of such an outcome is progressively becoming more apparent, especially in regard to the climate crisis which can subvert a plethora of natural processes. Besides, military training outside of active combat significantly contributes to the exacerbation of climate change and, if not alleviated before reaching the threshold, would become an irreversible development with catastrophic effects. Seemingly enough, there may be no proportional rationale to justify either the act of entering a war or the range of conduct during one due to the potential of extreme environmental costs of modern warfare. This concern is only furthered by international law offering limited environmental protection, since there isn't the necessary degree of safeguards to guarantee that conflict will result in inconsequential harm to the environment. Therefore, if the Just War Theory is intended to be normative guidance to international law, then the promotion of peace must strictly be upheld through means with no or minimal environmental costs to preserve and prioritize Earth's health.

5 Conflict Pollution of Air, Water, and Soil Resources

Conflict pollution negatively impacts the quality of air, water, and soil resources during times of active combat as well as after its cessation. These three fundamental and interconnected components of the ecosphere are responsible for many processes within environmental sustainability, such as the regulation of climate, distribution of sustenance for all species, and support for chemical reactions in ecosystems. According to the annual global climate report, 2024 set pace to overthrow 2023 as the warmest year recorded in the National Oceanic and Atmospheric Association's (NOAA) history spanning the last 175 years. [8] This record-setting heat shouldn't be taken lightly as its consequences range from extensive drought, frequent wildfires, extreme weather phenomena, loss of biodiversity, and rising sea levels along with the displacement of countless species. The additional air pollution of conflict-related greenhouse gas emissions, stemming from "damaged infrastructure, the loss of vegetation, and delivering humanitarian aid," aggravate the warming process and increase the risk of exceeding the climate threshold. [9] In reference to water sources, there is direct pollution from military vessels along with their equipment contaminating bodies of water and indirect pollution from the spilling of pollutants due attacks on industrial areas. Considering the scarcity of fresh water sources, which are vital to life on Earth, conflict is likely to weaponize this fact and is capable of endangering biodiversity in the process. Likewise, war is an instigator of soil contamination, which also impacts groundwater, through a wide array of toxic elements like heavy metals in ammunition casings and even radionuclides from nuclear explosives. Since the soil is responsible for support of vegetative growth and agriculture, pollutants from conflict can poison food supplies and compromise soil integrity through chemical disturbances.

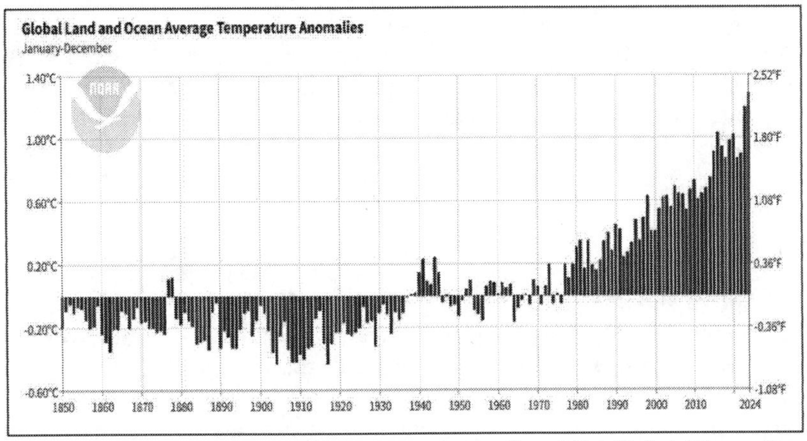

Fig. 1. Global Land and Ocean Average Temperature Anomalies, 2024 Global Climate Report [8]

Excluding other human activities that exacerbate climate change, such as burning fossil fuels to generate energy and electricity for infrastructural purposes, engaging in conflict is capable of catastrophic ozone depletion. More specifically, the recent developments in international affairs are an area of great concern given the pre-existing status of the environment along with the onset of simultaneous wars between multiple nations. This would become especially alarming if these current conflicts transpire to a scale similar to previous world wars and or involve the deployment of nuclear weapons. Though initiatives are in place to combat the climate crisis and improve air quality, they're typically established in reference to the analytics of social systems in a relatively non-warring world. As one of the most prominent consumers of fuels, global militaries already produce "5.5% of all greenhouse gas emissions" and, in comparison to country rankings, have the fourth-highest carbon footprint after China, the United States, and India respectively. [10] Besides, military emissions figures are predisposed to lacking accuracy due to incomplete data, unreliable reporting, and long-established exemptions within major international agreements. An example of this is the Paris Agreement, in which participating nations aren't mandated to collect or report statistics on emissions from military buildings, vehicles, ships, aircrafts, weaponry, and other equipment. Even if the treaty recommends voluntary reporting that provides relevant information to some extent, the fact remains there is no systematic method to gauge the trajectory and influence of aggregate military emissions on climate change at this time.

Additionally, there seems to be a rise in the effects of conflict pollution on water resources in modern warfare due the increasing scarcity of freshwater and inherent reliance on hydration to sustain life. In reference to the Israeli-Palestinian conflict, "during the May 2021 escalation, 290 water infrastructure objects were damaged, inflicting $10–15 million in damages", which generated risks for cross-contamination with wastewater for the citizens and other species in proximity. [11] Water pollution is also capable of disturbing natural aquatic systems through introducing harmful concentrations of toxic chemicals, such as heavy metals, either through active combat or the operations of military bases outside of war. A notable example would be the historic use of perfluoroalkyl and polyfluoroalkyl substances (PFAS) in several forms from jet fuel to aqueous film-forming foam (AFFF) for firefighting, which led to militaries becoming one of the leading polluters of PFAS in the world today.

[12] Regardless of whether it be from training, emergency services, or the onset of war, PFAS are capable of contaminating as well as altering the pH of groundwater sources and, in turn, drinking water. Furthermore, PFAS are among the most durable synthetic chemicals as they resist breakdown in the environment and human body, thus causing long-term disruptions. [12] Similarly, the underwater decomposition of military equipment, weaponry, ammunition, and sunken modes of transport release chemical compounds, fuel residues, and other pollutants into bodies of water. Once submerged the removal of these objects is not only complicated but also costly, which typically results in them remaining underwater indefinitely and being active sources of pollution for extended durations of time. Aside from the unintentional release of pollutants from the aforementioned improper disposal, there is even the direct weaponization of chemical warfare agents to pollute water resources. As an example, the Baltic Sea still contains contaminants, such as mustard gas, from chemical munitions dumped into the water during World War II that negatively impact the environmental quality of sediment, vegetation, and marine life due to their corrosive properties. [13]

Considering most combat is fought on land, soil resources are markedly exposed to activities of warfare and their corresponding pollutants. The "warfare-induced disturbances to soil are basically of three types – physical, chemical, and biological", which compromise soil quality to become unsuitable for both natural and anthropogenic processes. [14] Physical disturbances involve disruption from erosion or compaction of soil horizons, which can result from military vehicle traffic, troop movement, or underground excavation for tunnels and trenches. If the structure of the soil is fundamentally modified, especially in an abrupt fashion, then plant life as well as soil biota will suffer from environmental stress that jeopardizes soil health by obstructing nutrient cycling. On the other hand, chemical disturbances influence the composition of soil by embedding harmful chemical compounds that reduce fertility and problematize the soil's capacity to support life in underground ecosystems. Whether it be from ammunition or other military debris, extensive deposits of lead, zinc, and arsenic among other chemicals prompt shifts in pH from fluctuations in soil toxicity and heavy metal enrichment. [14] At that, it's possible for warfare to stimulate multiple types of disturbances at the same time, such as bombturbation in which explosives change physical topography by cratering the surface while depositing nitroaromatic contaminants like Trinitrotoluene (TNT) into the soil composition. Likewise, biological disturbances, aside from the deliberate introduction of hazardous microorganisms like anthrax from Bacillus Anthracis bacteria, are commonly a resultant of physical and chemical disturbances that over time exhibit negative repercussions on organisms that reside in soil. An extreme but fathomable example in modern times would be the employment of nuclear weapons, since an event of radioactive fallout from nuclear devices can bioaccumulate within the food chain for decades and eventually destabilize surrounding ecosystems. Moreover, the possibility of nuclear warfare incorporates the physical disturbance of a massive explosion, the chemical disturbance of radiation, and the biological disturbance of extreme biodiversity loss in the afflicted area all at once.

6 Policy Concerns and Recommendations

Given the rate of globalization in the contemporary period, international negotiations and agreements are more so equipped to address and regulate conflict to the degree necessary for environmental sustainability. International organizations and laws are becoming increasingly essential for mediation in an interconnected world of modern technologies, since the advent of internet and global telecommunication networks prompt correspondence in ways once not possible. These newfound interactions and circulations reform foreign policy, since global connectivity has implications for the extent of exchanges between nations and broadening the likelihood of possible disputes. Though there are numerous organizations active in international conservation, the United Nations (UN) is one of the leading authorities for the global agenda on environmental policy. Even if current contributions, such as the legal principles outlined in the Protection of the Environment in Relation to Armed Conflicts (PERAC), are valuable to the prevention of conflict pollution to some degree, modern warfare is still a major hazard to the environment. Evidently, despite the prohibitions within international laws, recent wars are causing grievous environmental damage under the guise of conflict with little to no corrective measures. The Russo-Ukrainian conflict is a prime example as it already released "100 million tonnes of carbon" during combat and issued damage to Nord Stream pipelines causing "the biggest ever point source release of methane." [15] This is a crucial setback in collective measures to counteract emission output of greenhouse gasses and stop the warming process before global temperatures surpass the climate threshold. As an international organization with environmental responsibility, setting precedence for amenable regulations is imperative to respond to ongoing variations in atmospheric composition. Furthermore, such acts shouldn't be permissible nor go without rigorous penalization, even in instances of conflict, when considering the increasing predominance of environmental vulnerability.

At that, there is still a possibility of the weaponization of energy supplies at the expense of prior global efforts in the preservation of natural resources, which technically doesn't violate the proscription on weaponizing the natural environment itself. This is an intentional strategy to destabilize a country's viable sources of energy and electricity, which are necessary for the maintenance of internal systems that support and mobilize citizens, with the environment then being a casualty of association. As an example, the Russian invasion methodically targeted Ukraine's energy infrastructure and power grid as a means to quell their agency as a sovereign nation to confront threats within its borders. Consequently, this stimulated immediate concern for Ukraine due to inducing a collapse of renewable alternatives, including "90% of the country's wind power and 50% of its solar energy capacity taken off-line" from conflict. [15] Regardless of whether it be a war between two countries or a world war, there will be implications on global initiatives for environmental security risks as well as the creation of novel sources of pollution that inhibit our ability to uphold planetary health. Hence, the UN alongside other organizations should work towards policy-making that establishes comprehensive coverage for environmental protection, from barring the indirect weaponization of the environment to ensuring materials for military use are free of withstanding chemicals. Moreover, militaries shouldn't be considered exceptions in international arrangements to mandatory accountability for the various forms of pollution caused from their activities. As previously mentioned, the Paris Agreement only entails voluntary reporting of emissions by militaries, which is necessary information to better understand human input relative to the climate threshold and devise strategies to align with its natural capacity. If anything, the environment has consistently played a secondary role to human society in most regards, including conflict, but that approach isn't sustainable when on the verge of irreversible damage and uninhabitability.

If environmental sustainability requires human dilemmas to be considered a peripheral concern, then there are methods of issuing priority to planetary health that can be applied to international organizations. To start, international conventions and debates need to center dialogue of conflict around environmental protection as a non-negotiable, which ought to promote laws that interpet said environmental costs as paramount to humanity. If the narrative that the causes of war are just as or more pressing than the effects of war continues, then the environment will suffer degradation to the point of ecosystem destabilization and collapse. The UN must also continue to reinforce diplomacy and other peacekeeping operations, such as its multilateral treaties for disarmament, since war as an avenue to peace can inevitably induce Earth's uninhabitability and undermine any existence of peace. Similar to how the Just War Theory adapted to variations in circumstance throughout several time periods, the same should be true for the international regulation of war in an era of unprecedented modernization and technological innovation. Moving forward a just cause and right intention from a legitimate authority shouldn't be enough to justify conflict, as the potential of environmental catastrophe would endanger the ability to achieve peace for every living being on the global scale. Advancements in military tactics, automation of machinery, and the proliferation of weapons of mass destruction create harrowing circumstances for modern warfare potential, especially compounded with a state of environmental vulnerability. International laws must be attentive to threshold capacities, actively evaluate environmental trends, and frame the justification of war in proportion with the environmental costs the Earth can afford. In short, policy-making for environmental sustainability relative to conflict pollution ought to necessitate peace not only being extended to people but to the planet as an entity in its own right.

7 Conclusion

If an objective of foreign policy is the active promotion of global interests, then we must simultaneously function as advocates for the land and space in which it materializes. There has been one casualty present for and impacted by every outbreak of conflict in history, and that is the environment. In turn, it ought to be of our intention to be heedful of the role anthropogenic activities, including the inception of conflict, have in the planet's livelihood and habitability. Since we must necessitate environmental protection and pollution prevention at the forefront of our agenda to ensure planetary health, there ought to be more consideration for whether conflict is justifiable when the environment is in a position of increasing susceptibility to human activity and culture. Though it may be impossible to avoid wars altogether, I argue there is a pressing need to amend international laws to have more strict regulations on the environmental modifications and subsequent costs of conflict. As exhibited in recents wars, there are methods of warfare that aren't explicitly prohibited by international agreements that still cause pervasive and potentially irreversible damage to the environment. If it's of our collective benefit to promote both sustainability and proportionality, then there should be a reassessment of standards for the legitimization of war in respect to the amount of conflict the environment is able to support without exceeding threshold capacity. Seemingly enough, current policy frames justification in terms of what circumstance, such as social atrocities or ideological pursuits, the conflict is in response to while not assigning greater precedence to what environmental costs Earth can physically bear. There instead needs to be emphasis on whether conflict is still a viable solution in an era of substantial environmental fragility that requires the utmost attention and timely strategization for countermeasures. Withal, this isn't necessarily to say that war is a luxury, as it may very well prove to be an unavoidable aspect of human existence, but a call to endorse alternatives with less environmental costs and more exhaustive regulatory practice for environmental protection within international law.

References

1. United Nations (UN). Chapter VI protection of the environment in relation to armed conflicts, International Law Commission (2019)
2. Ochsmann, W., Grasshoff, M., & M, Gudo. The early evolution of the planet Earth and the origin of life, Palaeobiodiversity and Palaeoenvironments, **82**(1):284-294 (2002)
3. Groffman, P. M., , Baron, J. S., Blett, T., Gold, A. J., Goodman, I., Gunderson, L. H., Levinson, B. M., Palmer, M. A., Paerl, H. W., Peterson, G. D., Poff, N. L., Rejeski, D. W., Reynolds, J. F., Turner, M. G., Weathers, K. C., & Wiens, J. Ecological thresholds: the key to successful environmental management or an important concept with no practical application? Ecosystems, **9**(1):1-13 (2006)
4. Vision of Humanity (VOH). Global peace index: highest number of countries engaged in conflict since World War II, the Institute for Economics & Peace (2024)
5. Thorpe, N. Anthropology, archaeology, and the origin of warfare, World Archaeology, **35**(1):145-165 (2003)
6. Johnson, J. T. The just war idea: the state of the question, Social Philosophy and Policy, **23** (1):167-195 (2006)
7. Nicolaides, A. Waging a just and ethical war – contemplating Saint Augustine, 'just war' theory, Gaza, and other philosophical notions, Pharos Journal of Theology (2024)
8. National Oceanic and Atmospheric Administration (NOAA). Annual global climate report, National Centers for Environmental Information (2024)
9. Darbyshire, E. & Weir, D. How does war contribute to climate change, Conflict and Environment Observatory (2021)
10. Kinney, E. Time to face up to the military blind spot in global carbon accounting, Conflict and Environment Observatory (2022)
11. Hall, N., Kirschenbaum, A., and Michel, D. The siege of Gaza's water, the Center for Strategic & International Studies (2024)
12. Environmental Protection Agency (EPA). EPA and U.S. army announce joint sampling project to identify PFAS contamination near army installations (2024)
13. Vanninen, P., Östin, A., Bełdowski, J., Pedersen, E. A., Söderström, M., Szubska, M., Grabowski, M., Siedlewicz, G., Czub, M., Popiel, S., Nawała, J., Dziedzic, D., Jakacki, J., & Pączek, B. Exposure status of sea-dumped chemical warfare agents in the Baltic Sea, Marine Environmental Research, **161**:105112 (2020)
14. Certini, G., Scalenghe, R. & Woods, W. The impact of warfare on the soil environment, Earth-Science Reviews, **127**:1–15 (2013)
15. Brown, O. How Russia's war on Ukraine is threatening climate security, the Royal Institute of International Affairs (2023)

Synthesis of Nitrogen Slow Released Fertilizer Based Biochar Derived from Durian Shell

Chattreeya Booncharern[1,a], *Teerat* Boontha[1,b], *Kulthida* Thongnam[1,c] and *Duangkanok* Tanangteerapong[2,*]

[1]Demonstration School of Khon Kaen University, Thailand
[2]Department of Chemical Engineering, Faculty of Engineering, Khon Kaen University, Thailand

Abstract. This study aims to synthesize slow-release fertilizer using biochar derived from durian shell which is the main contributor to agricultural waste and become an environmental problem. Therefore, the utilization of the durian peel was carried out. Biochar was synthesized through the pyrolysis process at 450, 500 and 550°C, then impregnated with urea fertilizer and encapsulated with a biopolymer. Nutrient release rate of biochar coated with chitosan by dissolution method was studied. The result showed that yield of biochar decreased with increasing pyrolysis temperature. The BET showed that 500°C gives the highest surface area and pore volume at 0.34 m^2/g and 1.86 x 10-3 cm^3/g. The FT-IR analysis indicated that N_2 impregnated fertilizer had higher nitrogen amount resulting from amine functional groups where the peak at 3456 cm^{-1} was observed. This confirms that high porosity biochar can absorb urea fertilizer. There was some increase in functional groups representing chitosan after encapsulation, at 1156, 1453, and 1682 cm^{-1}, indicating chitosan was successfully coated on the biochar fertilizers. Nitrogen release results showed that synthetic biochar fertilizer can slow down the release of nitrogen by 60%. In conclusion, durian shell can be used to synthesize biochar and slow released fertilizer which prolong the release of the nutrient.

1 Introduction

Nowadays, Chemical fertilizers are widely used in agriculture. As they can release nutrients that plants need and make it easy to use nutrients. However, the fast release of chemical fertilizers can cause a loss due to leaching or evaporation. This usually happens before the plants receive nutrients. This can lead to insufficient nutrient in plants, requiring farmers to use more fertilizer, causing waste and negative environmental impacts. Therefore, slow released fertilizers have been invented to solve these problems. The slow released fertilizers contain nutrients that are absorbed by biochar and coated with polymer to slow down the release. Biochar is carbon-based material with high porosity. It can be produced by heating biomass in the absence of oxygen or by pyrolysis process. Due to its porous surface, it has been used as a nutrient absorbent in slow release fertilizers [1, 2].

* Corresponding author: duangkanok@kku.ac.th

Durian is economic fruit in Thailand. The amount of durian shell left over from the processing process is up to 75-80% of the output, Causing a large amount of durian peel to be wasted each year. It was found that durian peel has Lignocellulose containing 57 - 64% cellulose, 30.7% hemicellulose and 13.6% lignin. From durian peel biochar analysis, It was found the organic material 41.10%. There were important nutrients for plants, including 1,563 milligrams of phosphorus per kilogram and contains 47,500 milligrams of potassium per kilogram. From the above mentioned features, the synthesis of slow release fertilizers based biochar derived from durian shell are carried out in this research as a guideline for utilization of these waste materials.

2 Materials and Methods

2.1 Synthesis of Biochar from Durian Shell (DSB)

Durian shell (DS) was chopped into pieces and oven dried at 60 °C for 24 hours, then ground into 35 mesh (0.5 millimeters) sieve size. DS was pyrolyzed by Vulcan muffle furnace model 3-550 at 450, 500 and 550 °C using the heat rate of 10 °C/min for 60 min.

2.2 Preparation of Urea Infusion into Biochar and Making of Slow Release Fertilizer

Preparation of the durian shell biochar doped nitrogen (DSBN) was performed by adding 1 gram of urea fertilization (URF) mixed with 1 gram of biochar and 20 mL of water in a beaker and stirring on a hot plate at 120 rpm for 2 hours. The beaker was covered with the parafilm to prevent the evaporation, then DSBN was dried at 65 °C for 24 hours and coated DSBN with chitosan [3]. Chitosan solution was prepared by dissolving 1 g of chitosan and 100 mL of 2% (v/v) acetic acid. After that, DSBN was impregnated in 25 mL prepared chitosan solution, stirred in a hot plate at 120 rpm for 2 hours, then dried at 75 °C for 24 hours to obtain the biochar fertilizer coated chitosan (DSBNC).

2.3 Structural and Chemical Analysis

Surface area and porosity analysis of DSB at 450, 500 and 550 °C was examined by BET measurements and the functional group of DSB, DSBN and DSBNC was analysed using Fourier Transform InfraRed spectroscopy (FTIR).

2.4 Nutrient Release Test

1 gram of DSBN, DSBNC and URF was mixed individually with 100 mL water in different conical flask and covered with parafilm to prevent evaporation of fertilizers. All samples were stirred by a orbital shaker at 100 rpm for 10 days, and after that, 10 mL of each samples was collected every day for nitrogen release analysis using the Ultraviolet–visible spectroscopy (UV-Vis) [4].

3 Results and Discussions

3.1 Yield Analysis

Fig. 1 shows the yield of durian shell biochar produced from the pyrolysis at 450, 500 and 550 °C. The results showed that increasing the pyrolysis temperature from 450 to 550 °C reduced in the yield of durian shell biochar (DSB) by 20%. This might be due to the destruction of some compounds, such as cellulose and hemicellulose, during the pyrolysis process [5].

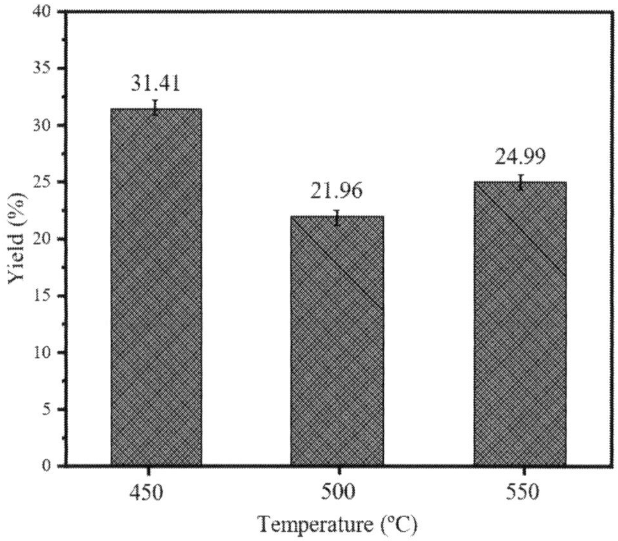

Fig. 1. Yield of durian shell biomass (DSB) from pyrolysis at temperatures.

3.2 Surface Area and Porosity Analysis

Table 1 shows the specific surface area first increased and then decreased with increasing pyrolysis temperature. The BET measurements showed biochar from pyrolysis process at 500 °C had the highest surface area and pore volume at 0.51 m^2/g and 1.86 x 10^{-3} cm^3/g. The decrease of surface area and pore volume after 500 °C could be due to the destruction of the pore structure caused by secondary pyrolysis [6]. The average pore size on the biochar surface is 7.89 nm, which is in the range of 2-50 nm. Therefore, it is classified as a mesoporous material.

Table 1. BET Surface area and porosity analysis of DSB.

Temperature (°c)	BET Surface area (m^2/g)	Total pore volume ($\times 10^{-3}$ cm^3/g)	Micropore volume ($\times 10^{-3}$ cm^3/g)	Average pore diameter (nm)
450	0.29	1.46	1.24	4.63
500	0.51	1.86	1.10	7.89
550	0.20	1.76	1.58	4.81

3.3 Functional Group Analysis

The FTIR spectrum of durian shell (DS) showed a sharp peak at 1028 cm^{-1} corresponding to the C-O-C stretching of hemicellulose ester. After pyrolysis at 500 °C, the peak at 1028 cm^{-1} disappeared in all biochars, indicating that hemicellulose is decomposed. There was also a broad peak of O-H stretching around 3400 cm^{-1}, which is related to alcohol from cellulose or phenols from lignin. The weak absorption band at about 1600 cm^{-1} is assigned as the C=C bond in the aromatic compounds. After the DSB was impregnated with fertilizer, the strong absorption peaks at 3213 to 3456 cm^{-1} were ascribed to O-H and N-H stretching, confirming that biochar can absorb urea fertilizer (DSBN). There were some increase in functional groups representing chitosan after encapsulation (DSBNC), namely amine asymmetric stretching, amine bond, and secondary amine at 1156, 1453 and 1682 cm^{-1}, respectively. The absorption peaks slightly shifted after coating with chitosan, suggesting that there are chemical interactions occurred on biochar surface.

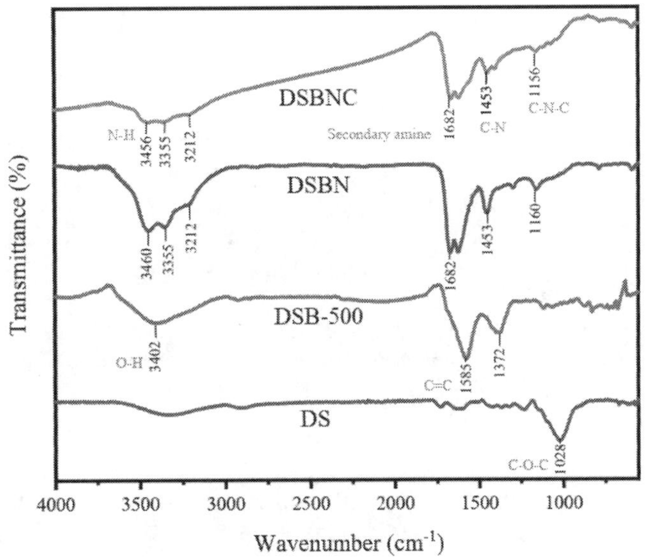

Fig. 2. FTIR spectrum of DS, DSB at 500 °C, DSBN and DSBNC.

3.4 Nutrients Release Analysis

The measurement of nitrogen release by UV-Vis in water for 10 days is shown in Fig 3. It can be noticed that the maximum release of encapsulated fertilizer (DSBN and DSBNC) appeared to be 40% and that of uncoated fertilzer is 100% after dispersing for 1 day. However, after prolonging for 10 days, DSBNC and DSBN showed the similar trend in nitrogen releasing in water. The findings of this figure showed that the synthetic biochar fertilizer both DSBN and DSBNC can significantly slow down the release of nitrogen by 60%when compared with the commercial urea fertilizer (URF). This has been seen in the case of urea-coated fertilizer improved oil palm growth due to a higher rate of volatilization in dry region [7]. Therefore, coating the durian shell biomass with chitosan dis not show any significant differences in nitrogen release.

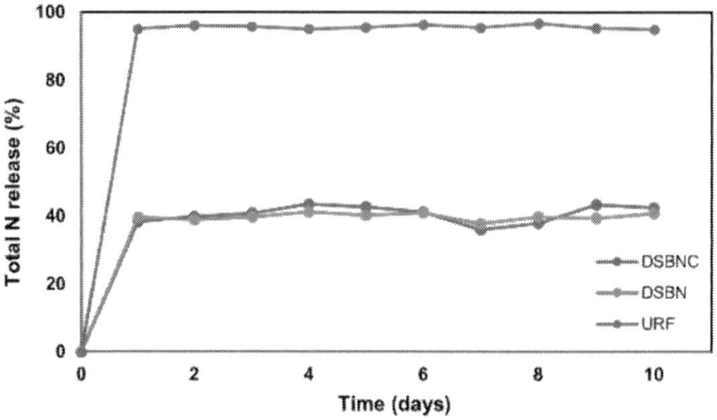

Fig. 3. Total nitrogen release of URF, DSBN and DSBNC.

4 Conclusion

Durian shell can be utilized as a viable biomass source to synthesize biochar and produce a slow-release fertilizer. The FTIR results showed the discovery of functional groups that represent urea and chitosan, confirming that DSB can be used as an adsorbent in slow-release fertilizers. From the release test, the measurement of nitrogen release by UV-Vis showed that DSBNC can slow down the release of nitrogen compared to the standard urea fertilizer and thus, could be a promising fertilizer in the agricultural industry.

This project was supported by Science Classroom in University Affiliated School (SCiUS). The funding of SCiUS is provided by Ministry of Higher Education, Science, Research and Innovation, Thailand.

References

1. Thonglem, S. and Intawin, P. Characterization of biochar derived from durian shells by pyrolysis process. Journal of RMUTI, 2020; 3: 44-56.
2. Wang C, Luo D, Zhang X, Huang R, Cao Y, Liu G, et al. Biochar-based slow-release of fertilizers for sustainable agriculture: A mini review. Environmental Science and Ecotechnology. 2022;10: 100167.
3. Gumelar DM, Hamzah M, Hidayat SA, Saputra AD, Idvan. Utilization of Chitosan as Coating Material in Making NPK Slow Release Fertilizer. Macromolecular Symposia. 2020;391: 1900188.
4. Trinh HT, Dao NQ, KuShaari K, Ismail L, Shuib A. Study the release of urea from Agrium coated urea using UV-Vis spectrometer, Applied Mechanics and Materials. 2014;625: 745-8
5. Al-Wabel MI, Al-Omran A, El-Naggar AH, Nadeem M, Usman ARA. Pyrolysis temperature induced changes in characteristics and chemical composition of biochar produced from conocarpus wastes.Bioresource Technology. 2013 Mar 1;131:374–9.
6. Liao W, Zhang X, Ke S, Shao J, Yang H, Zhang S, et al. Effect of different biomass species and pyrolysis temperatures on heavy metal adsorption, stability and economy of biochar. Industrial Crops and Products. 2022 Oct 15;186:115238.

https://doi.org/10.1051/e3sconf/202562005001

7. Rasid MN, Chek TC, Redzuan AF. Effectiveness of urea-coated fertilizer on young immature oil palm growth. Journal of Advanced Agricultural Technologies. 2014 Jun 1; 56-59.

AUTHOR INDEX